Our Universe

big bang to black holes explained

by Joe Tyndall

isbn # 979-8-952316-02-7

text and graphics created using open source tools
LibreOffice, GIMP and Inkscape
plus Libre Caslon, Inter and STIX Two fonts
donations suggested to support these groups
3D illustrations by the author using Lightwave 3D
typos, spelling and grammar checked by Claude AI
special thanks to booklaunchers.com
cover artwork by 100covers.com

**information links in the text create a bibliography
and do not imply an endorsement of this book**

Chapters

Clickable links, bracketed **[ ]**, plus the tables, diagrams and formulas can be found at **[joetyndall.com/b2].**

Classic Physics Primer

The knowledge in this book could be taught to 6th graders. I mean the concepts, not the advanced math to solve problems. To begin, we survey physics up to 1900. Reviewing this is essential because newer science builds upon and uses older science. Seven classes at MIT covered this material and the required math. A thousand hours of education is condensed into eight pages. Most science programs for a general audience focus on personality stories rather than concepts. We do the opposite to examine a few major snowflakes on the iceberg. To master classic physics, students work out problems first solved years earlier. Relax and enjoy. No exams or problem sets found here.

Science definition #1, a body of knowledge, always a snapshot

Promising and popular ideas have turned out to be wrong. I suspect some conjectures summarized in this book may be discarded. But as of today, this is our best thinking about our universe, tersely explained.

Science definition #2, a method for finding truth

A "theory" in science must be falsifiable. It involves six steps.

1) Observation of facts that are verifiable and reproducible.
2) A pattern is inferred from similarities in the observations.
3) Zero evidence contradicts the pattern. Or the theory is false.
4) The pattern yields a rule predicting new facts.
5) Experiments are run to check these potential facts.
6) Results from new observations confirm the rule.

Step **#3** tells us, if any evidence, solidly verified, contradicts the pattern, the theory must be modified or scrapped. For 400 years since the rebirth of science, this six-step loop has run continuously to generate all the modern miracles most folks take for granted. To function, science needs the free flow of information. This includes freedom of thought, speech, press and association. Science has emerged multiple times and, except for its current run, was later snuffed out.

What, how and why questions

What answers give a description. How answers provide a mechanism. Why answers offer a reason. Our understanding of any given topic could be what, how or why. Pushing toward why is the goal.

Language and intent

Our universe is designed so the laws of physics can never be violated. This sentence is true. But note the intentionality it implies. This feature is built into languages and is impossible to avoid. People are self-aware. We do things for reasons. Language reflects this. Verbs project choice and intent. However, our universe evolves in accord with physical laws without a consciousness needed to move things around. Magic is not required. Also note the word evolves. Evolution means change over time. The only constant is that things change.

Knowledge is layered

Physics builds on math. Does math already exist? Do we discover it? Or do we invent it? Chemistry always obeys the laws of physics. Biology and geology conform to chemical principles. Laws from lower levels are never violated. All other disciplines layer on top of this.

Equations and units

6th graders have arithmetic. High school adds algebra, geometry and trig. College gives calculus, differential equations and so on. This book only needs the concept of an equation with variables (**x, y, m, a**), constants (**π, c, ħ, 2...**) and operators (**+, -, ×, /, ( ), ^, ln, sin...**). For example, Newton's second law **F=m × a** is a rule where force **F** equals mass **m** times acceleration **a**. Multiplications **×** is implied, so **F=ma.** Beginning science uses the metric **mks** system, **m**eters for distance, **k**ilograms for mass and **s**econds for time. Powers of 10 notation avoids long strings of zeros. 13,800,000,000 years, the age of our universe, is $\mathbf{13.8\times10^{9}}$. 0.0000000047 is $\mathbf{4.7\times10^{-9}}$. As with common factors, units can be crossed out. To read a baby physics book, 1235 pages equals **5** pages**/**~~day~~ times **7** ~~days~~**/** (per) ~~week~~ times **35.3** ~~weeks~~ (**=5×7×35.3**).

1609, science (re)emerges

Galileo published his observations of the planets and moons using a telescope. Kepler published two of his three planetary motion laws.

This experiment-theory duo proved the Earth-centered cosmos was wrong. Copernicus had earlier suggested a sun-centered model with circular orbits. This was closer to the truth but also wrong. The correct orbit is an ellipse, a fact Kepler deduced from years of Brahe's planet position data. These ideas contradicted religious dogma, provoking the wrath of authorities. Was this blasphemy or simply the truth?

Early science needed literacy, resources and leisure time

Galileo sold telescopes to rich people to see arriving ships earlier. Had he not, the church likely would have killed him. By uncovering how our universe works, science finds new, profitable products faster. In 1660, upper class nerds founded the Royal Society to talk tech. In 1687, Newton published **Principia**, the first physics textbook.

Newton's three laws, a what description

1) Objects remain in uniform motion unless acted upon by a force.
2) Force equals mass times acceleration, **F=ma.**
3) Every force generates an equal and opposite reaction.

Law **#1**, if no force **F** acts, continue straight line motion at the same velocity **v** forever. If the object stays at rest, **v=0.** Prior to Newton, people believed things stopped moving naturally. Not true. Air resistance and friction slow moving objects. In a vacuum, lead balls and feathers fall equally fast, a fact demonstrated in 1969 on the moon.

Law **#2** gives an equation **F=ma** to predict motion when a force **F** pushes or pulls an object. Mass **m** is an object's resistance to changes in motion. The bigger the mass, the less acceleration **a**, which is how quickly speed changes. This includes altering direction when a force is applied at an angle to the object's line of motion. The time **t** the force is applied determines how much velocity **v** changes. $\mathbf{v=v_0+at}$.

Law **#3**, two frictionless ice skaters with mass $\mathbf{m_1}$ and $\mathbf{m_2}$ are at rest, **v=0.** Momentum **p=mv** is mass times velocity. $\mathbf{p_1=p_2=0}$ at rest. Mass and weight are not the same. Weight is a force **F=mg** gravity **g** causes. When you push a boat on water you feel its mass. Lift the boat, you feel its weight. Our skaters push apart. As they move away in opposite directions, we now have $\mathbf{p_1=m_1v_1}$ and $\mathbf{p_2=m_2v_2}$, The smaller skater moves away faster. However, start to finish, $\mathbf{p_1=p_2}$ is always true.

Does quantum mechanics and general relativity overthrow Newton? No. They extend his physics. **GR** modifies "straight line" motion in a gravitational field. **QM** explains how mass works at a deeper level.

Using websites like **[the engineering toolbox]** that give tables of physical properties, Newton's laws calculate stresses to design skyscrapers, bridges, mechanical systems and so on with confidence. This assumes we considered all relevant factors and have not made mistakes doing calculations. When I design, I always prototype and test.

Generalizing to our universe

Doing simple algebra on law **#3**, $m_1v_1 - m_2v_2 = 0$. This means the center of mass for the two skaters remains at the same point before and after they push apart. Enlarging this to include every other motion, everywhere else in our universe, for all time, tells us the sum of all momentum in our universe is zero. For everything moving, other things move in other directions to balance this motion out. When you walk forward, you rotate the Earth backward by a tiny amount. In a similar manner, all angular momentum in our universe also sums to zero. Angular momentum is stored in things rotating around an axis. The axis could be an axel. Examples include wheels and planets spinning.

Thermodynamics

The second major discipline in physics is heat and energy. Examining chemical reactions, steam engines and distillation codified this. By 1800, textbooks were available. There are four laws.

0) If systems A and B are in thermal equilibrium and B and C are in equilibrium, then A and C are also in equilibrium.
1) Total energy remains constant.
2) Entropy (disorder) always increases.
3) You cannot cool anything to below absolute zero.

For law **#0**, what is thermal equilibrium? Imagine a perfectly insulated box. No heat, light or molecules enter or escape. We leave this long enough to reach constant temperature and pressure throughout. Turbulence has dissipated. This is thermal equilibrium. While molecules and photons, particles of light, do bounce around, there is no net flow of energy, heat or molecules, causing things to change.

For law **#1**, a barrier splits our box into two equal volumes. On one side, we have hot, compressed gas. The other side is a perfect vacuum, zero gas. Open a valve. Molecules flow from the pressurized half. We use this to spin a turbine to power a light. Over time, both sides settle into thermal equilibrium. The flow stops. The light goes out. How much energy is in the box before and after this? Half the energy and molecules are in each half. Both sides are at lower temperature and pressure. But the total energy in the box remains precisely the same.

For law **#2**, after our law **#1** experiment, both sides are at thermal equilibrium. Total molecules is the same start to finish but they are not moving as fast (temperature) and are half as tightly packed (pressure). The molecules are more disordered. Entropy increased. Think of entropy as energy divided by volume or by total particles in a system. Photons are particles. They are tiny pieces of light. At the start, there were none. Then we used energy to make photons. The number of molecules plus photons has increased. Same energy, more total particles. So the molecules have even less energy per molecule.

For law **#3**, pump everything from the box. It is completely dark and empty. What is the temperature inside the box? Absolute zero, -273.15 Celsius (-459.67 Fahrenheit). You can extract no more energy. However, slightly above absolute zero, some materials exhibit superconductivity and superfluidity, both actively studied for deeper insights.

The gas law

PV=nRT equates pressure **P**, volume **V**, total atoms or molecules **n** in moles ($\mathbf{6.02 \times 10^{23}}$ particles), the gas constant **R**, which equals 8.314, and temperature **T**, measured in degrees Kelvin. Kelvin is the same scale as Celsius except zero is at absolute zero, not where water freezes.

What is energy?

Energy is an accounting tool. We always have the same amount. Imagine a box around a system of interest. If energy passes in or out, you must consider it. The sun warms the Earth. Life eats food. Enlarge your box to include all energy sources and drains. When energy flows from places of higher concentration to places of lower concentration, change happens. Things move, heat up or cool down. Heat flows by conduction, radiation and convection. There are formulas to compute how much energy we can use from wind, light, falling water, chemical bonds (burning stuff), nuclear power, batteries, motion, heat, weights in the air, wheels spinning and so on. We apply energy to do work, meaning to change things. Energy has quality. Energy as electricity is more easily applied. For classic physics, law **#2** says no energy exchange is 100 percent efficient. Some energy always converts to useless low temperature heat. However, in particle physics, interactions can be perfectly efficient. We now have (**3+4=**) **7** laws of physics.

Electricity and magnetism

By 1900, work on **E&M**, the third major classic area, had been unified into four equations. Now we have (**3+4+4=**) **11** physical laws. Shortcut laws to simplify and summarize can be derived from these eleven. Charge q comes as positive (+) for protons and negative (-) for electrons. Like charges repel. Opposite charges attract. In the mks system, the force between two charges is $F=kq_1q_2/d^2$ where k is 8.99×10^9, q is in coulombs, 6.24×10^{-18} electrons, and d is meters between the charges. An electron has a mass of 9.11×10^{-31} kilograms. At the size of an atom, 10^{-10} meters, the force from charge on an electron from a proton is 2.54×10^{-8} newtons. The gravity force $F_g=m_1m_2G/d^2$ is 1.02×10^{-47}. Gravity is 2.5×10^{39} times weaker than electromagnetism.

Vectors and fields

Vectors are imaginary arrows having a magnitude and direction. For example, driving 60 miles per hour to the northeast. Vectors are used extensively to solve problems. **E&M** also uses fields that assign a vector to every point in space. The math handles this nicely.

Maxwell's 4 laws

Law **#1**, $\nabla\cdot E=\rho/\varepsilon$, tells us a charge distribution ρ causes an electric field **E.** ε is a constant that depends on the material in which the field is embedded. If we have fields from multiple charge sources, we can vector add these fields. Fields from different sources do not interfere with each other, a principle called superposition. If we place a test charge in a field, it moves, even though that charge and the charges causing the field(s) never touch. **E&M** field energy is stored in space itself. Remove the source charges instantly. It takes time for the test charge to notice. Empty space (nothing) is more complex than people believe. For a point charge source, doubling the distance (**2x**) from the source, reduces the field strength by a factor of four (down to **1/4**).

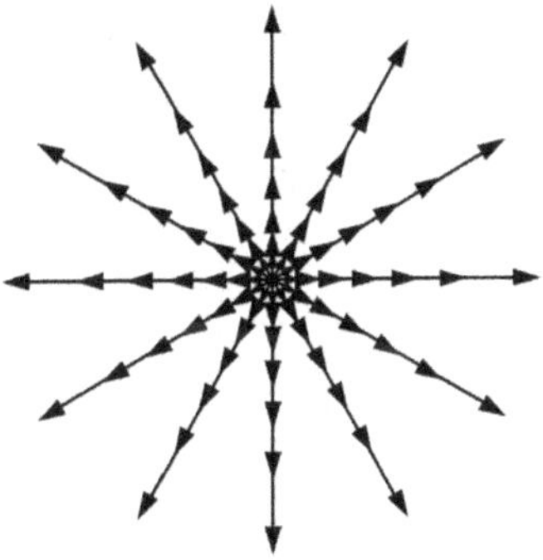

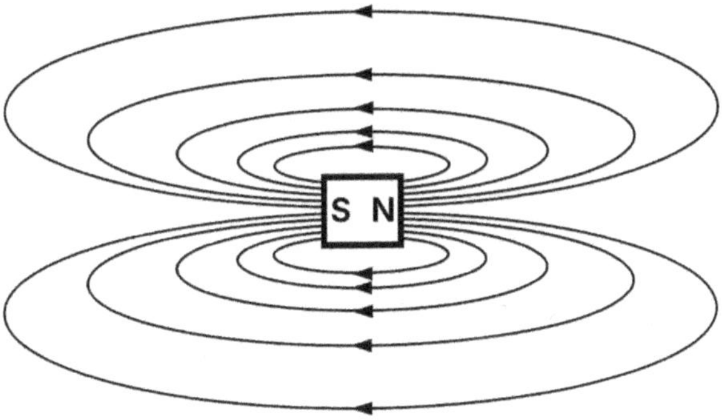

Law **#2, ∇•B=0**, says magnetic field lines **B** curve to form loops rather than extending straight out. Magnets with north and south poles placed in a magnetic field align then move along the curve. Again, the field strength diminishes with distance from the source. The right hand side is zero because there are no magnetic monopoles.

Law **#3, ∇×E=-∂B/∂t,** tells us a changing magnetic field **B** causes an electric field **E** that will move charges. This is the principle behind generators producing the voltage and current found in wall sockets.

Law **#4, ∇×B=μ(J+ε(∂E/∂t))**, says moving charges **J** or a changing electric field **E** cause a magnetic field **B** that can deliver force. This is how solenoids, transformers and electric motors work.

Combining these four equations explains how light waves travel through empty space. This also gives us the speed of light. **c=1** divided by the square root of **ε × μ**. Through empty space, this computes to 299,792,458 meters per second, 186,282 miles per second. **ε** and **μ** are properties of the material light passes through. Light travels slower in glass, water and other materials than through air or space.

Waves, Vibrations, Optics

Physics 8.03 used differential equations to solve increasingly complex problems, including reflection, refraction, interference, diffraction, polarization and transmission through space and other materials. This gave more derived shortcut laws but no new fundamental laws.

Getting to modern physics

By 1900, scientists had observed things classic physics could not explain. What is matter made from? Cut gold in half repeatedly. After each cut, is it still gold? Or, is there a smallest chunk that is no longer gold when divided? Radioactivity, blackbody radiation, the ultra-violet catastrophe and chemical spectra anomalies had been characterized (what type answers). How and why answers eluded physicists.

Determinism

If you perfectly measure where everything is and where it is going, classic physics suggests you can predict the future and likewise compute history. Was the destiny of our universe fully determined at the moment of the big bang? Do we actually have free will? We will defer this answer until we have looked at quantum mechanics.

Is there a limit to fundamental knowledge?

Knowledge comes as two types, detailed and fundamental. Detail is what you find at **[the engineering toolbox]**. Fundamental knowledge is how our universe is put together and operates. Look around your room. Except for consciousness, is anything you see not yet explained to some level by science? No. Will we finish our quest for fundamental knowledge? Thus far, folks claiming we have, or were near to this, were wrong. An opinion plus $20 buys a cup of coffee.

How we view our universe has evolved

The naked eye sees 10,000 stars. When language emerged, based on no evidence, and on fear, our ancestors created origin stories. The cosmos was small and newly created. These myths evolved. Some went extinct. New ones emerged. 12,000 years ago, we learned enough for permanent settlements. Around 1600, we invented telescopes and microscopes to view the realms of the very large and the very small. In 1900, telescopes still could not resolve stars in other galaxies. We thought our Milky Way was the entire universe, eternal and static. So, what has 20th century science and technology revealed?

Gravity & General Relativity

Gravity keeps us pinned to the Earth as we orbit the sun (a year) and spin on our axis (a day). Newton's gravity law $\mathbf{F_g=m_1m_2G/d^2}$ gives the force $\mathbf{F_g}$ between objects. The mass of the first $\mathbf{m_1}$ times the mass of the second $\mathbf{m_2}$ times a constant **G** ($\mathbf{6.67\times10^{-11}}$ m^3/kgs^2) divided by the distance **d** squared gives $\mathbf{F_g}$. This is the **(3+4+4+1=) 12**th classic fundamental law. It works extremely well for the weak gravitational field on Earth. But as telescopes improved, an anomaly was detected. And have you ever wondered how the Earth grabs your butt and yanks it down?

General relativity (GR)

Newton gave a what description for gravity. In 1915, Einstein showed us the how mechanism. His field equation $\mathbf{G_{\mu\nu}=kT_{\mu\nu}}$ is tensor calculus notation for a set of differential equations. **k** is a constant. The left side $\mathbf{G_{\mu\nu}}$ is the curvature of space-time. The right $\mathbf{T_{\mu\nu}}$ is the distribution of matter-energy. Matter-energy curves space-time. Curved space-time tells objects how to move. Gravity is geometry. Reading Einstein's 1916 **GR** paper, physicists soon realized only expanding or contracting solutions were possible. I have not shown Einstein's term added in 1917 to try to make our universe static. This antigravity force acts at intergalactic distances. In 1929, when we learned our universe is expanding, Einstein called inserting this factor his biggest blunder. Given what we know today, Einstein's mistake was, in fact, pure genius.

Einstein's first test

Big Al proposed three tests for **GR**. Theories in science must be falsifiable and must foretell new facts to observe. If any data found contradicts a prediction, and is verified, the theory must be scrapped or revised. To date, no indisputable evidence against **GR** has emerged.

Newton's gravity predicts every 87.969 days Mercury's orbit should return precisely to where it started. But this orbit precesses by a tiny amount. Over thousands of years, it traces out a daisy-like path. Pi (π),

3.14159265..., the circumference to diameter ratio, can be computed by the series **π = 4 - 4/3 + 4/5 - 4/7 + 4/9 - 4/11 +** (and so on). Einstein used a more complex infinite series to solve his math. The first term was Newton's gravity. The second predicts Mercury's precession. The third and later terms were not needed. **GR** is accurate. But we still use Newton to fly rockets to the planets. It yields an adequate answer and is much easier to use. We also send up fuel for course corrections.

Einstein's second test

Photograph stars. Then photograph the same stars during a solar eclipse. Compare the pictures. Stars that should be hidden behind the sun are seen because their light is bent by the sun's gravity. **GR** predicts the angle light bends. **GR** is now being used to weigh galaxies.

Einstein's third test

GR claims clocks at lower elevation (stronger gravity) tick slower than clocks at higher elevations. By comparing extremely accurate atomic clocks at different heights, **GR** has aced this test repeatedly.

Other GR validation

The global positioning system **GPS** needs precise timekeeping. An array of satellites broadcasts time and position. Cell phones intercept these signals then do the math to compute location. If the satellites do not correct for the **GR**-predicted change in how fast time flows due to altitude and velocity, **GPS** becomes worthless within minutes. Given a century of rigorous testing, we have high confidence **GR** is correct.

Four space-time length dimensions

Our universe has three dimensions of length plus time. To solve **GR** problems, multiply time **t** by the speed of light **c** to get **ct**. This gives a fourth dimension of length for the math. Is there a physical interpretation to this? Yes. Everything in our universe constantly travels through space-time (not space) at the speed of light. Weird, but true.

Tensors and calculus

Divide a curved surface into tiny squares. For every square, place a vector arrow, a tensor, in the middle perpendicular to its surface. Set the tensor length equal to the square's curvature. Convex square tensors point up. Concave square tensors point down. Tensor length for flat squares is zero. Making the squares smaller also makes them flatter. The tensor is shorter. In the limit, we have zero area divided by zero length. Divide zero by a number, we get zero. Divide a number by zero, we get infinity. But what is **0/0**? Calculus gives us that answer.

So how does gravity grab your butt and yank it down?

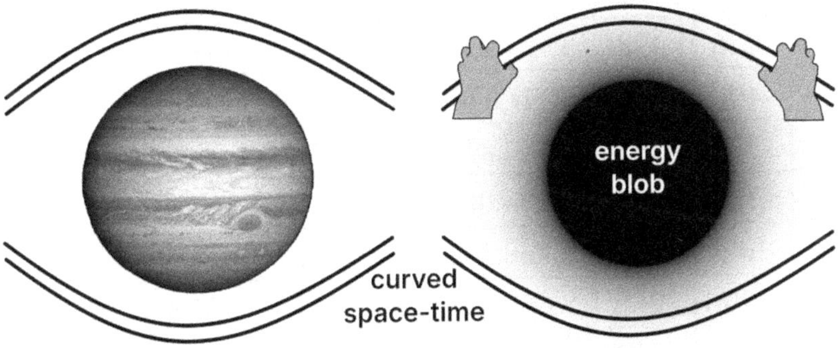

Let's say we have a planet. Its mass bends space-time near it. This causes a gravitational field, which causes objects to move toward the planet. Curved space-time is the gravitational field. Things move through space-time along the shortest possible path, which is also the path of least action. Least action is a deep physics principle still lacking how or why answers. Reverse this situation. Take a volume of space where gravity is weak. Space-time there is mildly curved. You float weightlessly. Next, grab space-time and bend it as much as it was bent by the planet. What happens? Space-time is the most rigid thing we have found. It takes enormous energy to deform it. Where does this energy go? It forms a planet, or more correctly, a big blob of energy. But energy is never found naked. It rapidly converts via $\mathbf{E=mc^2}$ into particles having mass. This occurs daily at particle accelerators like **[FermiLab]** and **[CERN]**. These particles then collapse into hydrogen atoms to create a gas planet. How big a planet? The blob's mass-energy exactly equals the energy used to bend the space-time.

A universe from nothing

If mass-energy $\mathbf{E_1}$ equals curved space-time energy $\mathbf{E_2}$, the planet and its gravity field have precisely zero net energy. $\mathbf{E_1-E_2=0}$. This is true for every object in our universe, be it a galaxy, golf ball or mote of dust. Summing all of this, our universe has zero total energy.

Switching Newton to Einstein

	sec	feet	ft/sec	GR feet (ct)	meters	m/sec	GR meters
A	1	-16	-32	983,568,960	-4.9	-9.8	299,791,809
B	2	-64	-64	1,967,137,920	-19.5	-19.5	599,583,619
C	3	-144	-96	2,950,706,880	-43.9	-29.3	899,375,428
D	4	-256	-128	3,934,275,840	-78.0	-39.0	1,199,167,238
E	5	-400	-160	4,917,844,800	-121.9	-48.8	1,498,959,047
F	6	-576	-192	5,901,413,760	-175.6	-58.5	1,798,750,856
G	7	-784	-224	6,884,982,720	-239.0	-68.3	2,098,542,666
H	8	-1024	-256	7,868,551,680	-312.1	-78.0	2,398,334,475
I	9	-1296	-288	8,852,120,640	-395.0	-87.8	2,698,126,285
J	10	-1600	-320	9,835,689,600	-487.7	-97.5	2,997,918,094

Drop a lead ball. How far does it fall over time? In 10 seconds, the ball drops 1600 feet. For a baby physics problem, we put 1 to 10 seconds in place of **A** to **J** in the diagram below. However, for **GR**, we multiply time by the speed of light to get **ct**. This gives us the **GR** feet column to use for **A** to **J**. As the ball drops 1600 feet, **GR** says it also travels 9,835,689,600 feet (1,862,824 miles) through space-time. This gives a feel for how gently the Earth's mass curves the space-time near us.

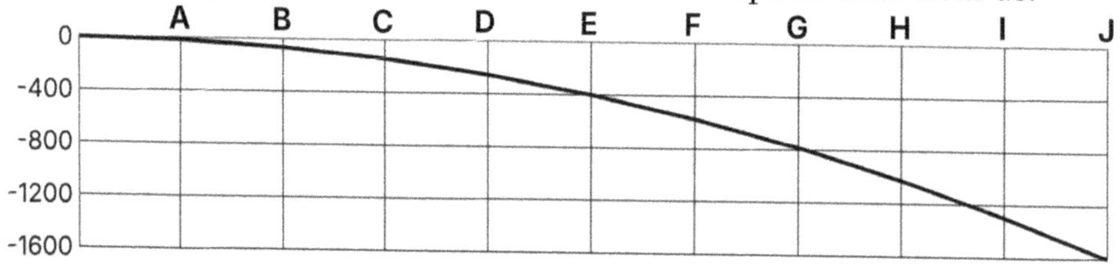

Space-time curvature causes gravity

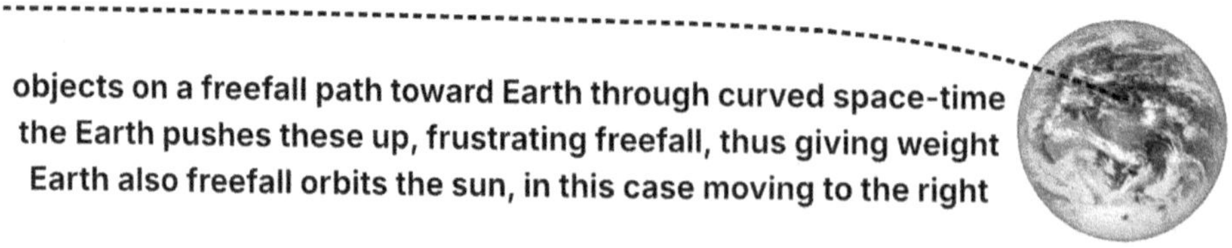

The sun curves space-time. Earth and some object orbit the sun. Our planet also curves space-time. So the object veers to Earth. But our planet blocks its path. Earth pushes things up, sometimes abruptly.

Parachuters are weightless to begin. After 10 seconds, they hit terminal velocity, 120 miles an hour, where air resistance equals acceleration from gravity. Deploying their chute, a larger drag, slows speed to 20 miles an hour. Every object travels at the speed of light through space-time along the straightest possible path. You are weightless until something pushes you up. This upward force is your weight. Note that objects moving through space-time flatten the curvature slightly as they move. Energy stored in space-time curves an object's path.

Making gravity

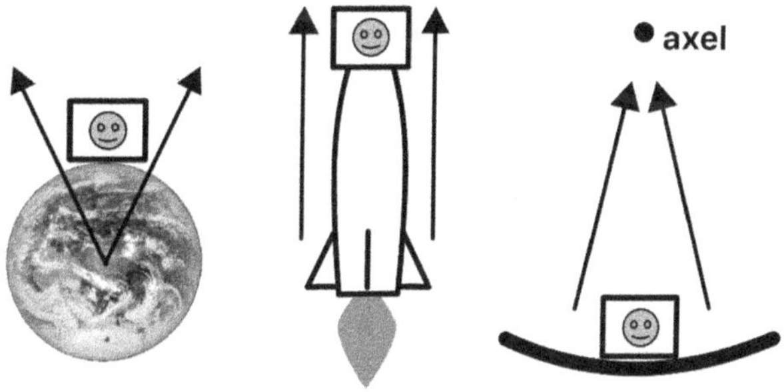

You are in a windowless room. How could you feel gravity? Put a planet under the room to bend space-time. Or, put your room on an accelerating rocket. Or, mount your room on the rim of a wheel spinning in space. Adjust these systems so your weight is exactly the same. Can you tell which situation caused the force you read on your bathroom scale? No. You must look outside. You could measure the angles the force acts at the edges of your room. But let's ignore this minor secondary effect.

The equivalence principle

Because you cannot tell how gravity acceleration was produced, the above three cases must be the same phenomena. Big Al knew this.

Gravity waves

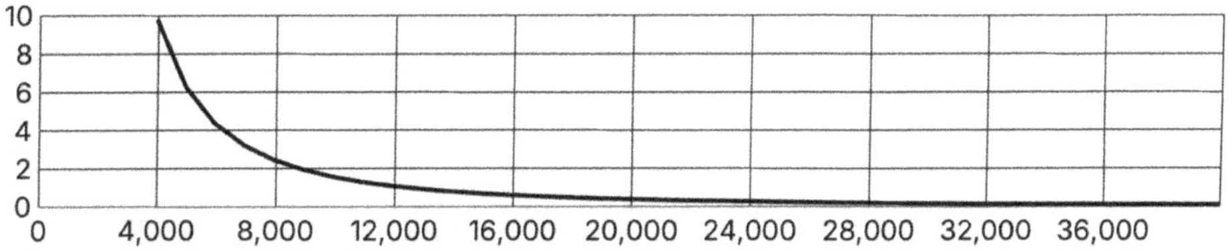

How does Earth signal that we have moved along our orbit around the sun? We send gravity waves, sapping a tiny bit of our orbital energy. **[LIGO]** has detected gravity waves from huge, distant objects spiraling into each other. Gravity wave strength diminishes with distance as the chart shows. At the Earth's surface, 3,963 miles from its center, gravitational acceleration is **-9.8** m/s^2 (the vertical axis). At the distance to the moon, 238,855 miles, gravity is only 0.0027 times as strong. Also, it takes time for our universe to learn things have changed. No mass, energy or information can travel faster than light.

GR sets the topology of our cosmos

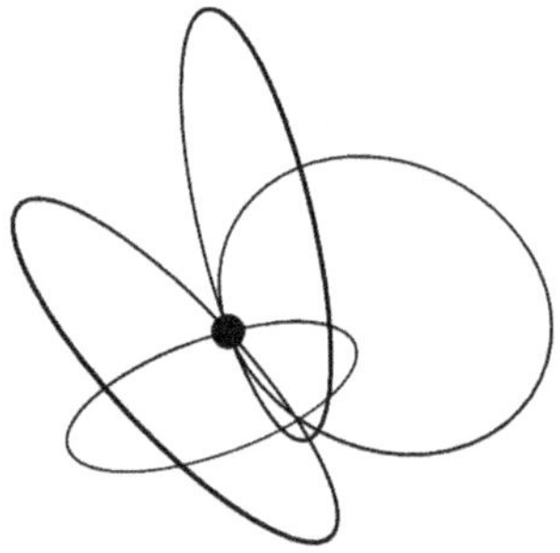

Pretend we have a telescope that can look infinitely far and fast and x-rays stars and dust. Gazing forward, what do we see? The back of our head. Next, aim it upward. What do we see? Again, the back of our head. No matter what direction we look, we always see the back of our head. Now, move the telescope to any other place in our universe. Again, we always see the back of our head. Every point in our universe is its center. Because our universe is expanding (proven) and because it is flat (proven), our universe is closed, unbounded and finite. Like the surface of a sphere, there is no edge. Because space-time is curved and closed in every direction, even traveling infinitely fast, we can never leave our universe.

What is outside?

Current thinking indicates outside our universe, space-time continues to inflate (more later). While **GR** tells us no mass or information can travel faster than light, there is no limitation on how fast space-time itself can expand. Our post-inflation bubble continues to grow at a moderate rate. Our universe is enormous but it is not infinite.

Common sense

Humans live on a planet where gravity holds a thin atmosphere. We orbit a typical star at a distance where our surface temperature allows liquid water. Air pressure is a mild 14.7 pounds per square inch. Speeds we experience are slow compared to light. Our air and magnetic field protect us from harmful radiation. Normal human lives exclude the very large and very small. Our thinking has adjusted to this. However, nothing guarantees that, at high temperature, pressure or velocity, intense radiation and enormous or tiny scales, our everyday notions should dictate physics. Prepare for more weirdness.

A thought experiment

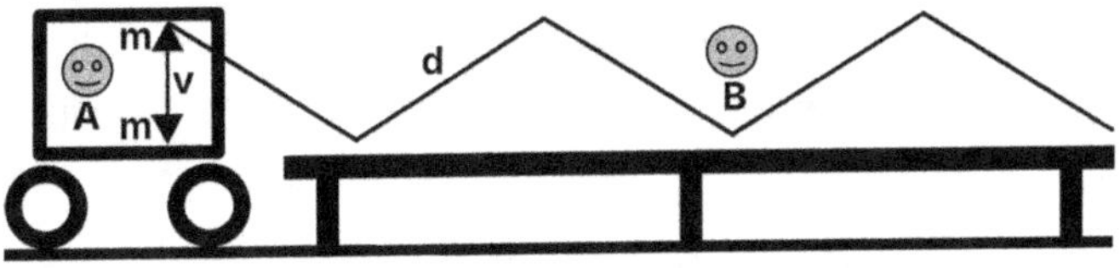

You **A** ride a train. Observer **B** on a platform watches you zoom by. You have a clock where you observe a beam of light bounce up and down vertically **v** between two mirrors. What does **B** see as the path of the light in your clock? **B** observes the zigzag diagonal **d.** But it is the same light. Maxwell's equations tell us all observers compute light speed to be the same, regardless of motion. In 1887, the Michelson-Morley experiment confirmed this. So, how can the up-down path **v** speed of light and the diagonal path **d** speed of light be the same?

Enter special relativity (SR) in 1905

SR tells us time slows as speed increases. If the train is smooth and at constant velocity, without looking outside, could you tell you were moving? No, exactly as with the source for gravity. For **A** inside the cabin, everything appears normal. But looking outside, **A** sees the entire panorama around them distorted into a circle ahead of them.

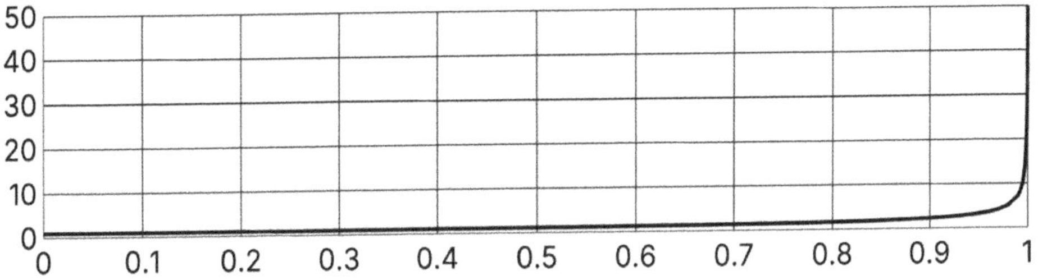

The horizontal scale is the fraction of light speed. The vertical axis is the factor by which time slows and mass increases. At low speeds, the curve stays near 1.0, no slowing and mass normal. At 90 percent light speed, clocks tick 43.6 percent as fast. At light speed, time stops. At human speeds, this effect is tiny. But at 99.99 percent the light speed, an astronaut's mass is 70.7 times what it was at rest. Increased mass explains why nothing with mass can reach light speed. Doing this requires infinite energy. Near light speed, a spaceship almost becomes its own black hole where time slows to a standstill. For a given speed, **SR** predicts the ratio of distance traveled to time experienced. At rest, it's all time **t**, no distance **d**. At light speed, it's all distance, no time. At intermediate speeds, $d^2+t^2=c^2$. You travel some distance and experience some time. Note, when velocity is constant, **GR** simplifies to **SR**.

The twin paradox

A paradox requires that you believe an untrue assumption. One twin goes on a journey around the observable universe at nearly the speed of light. The other stays home. Is **SR** symmetrical? No. The spaceship undergoes acceleration. Time slows nearly to a stop. If they accelerate at one **g** to remain comfortable, then at half way, decelerate at one **g**, the trip would take 57 years on the ship's clock. Returning home, their twin, who did not accelerate, died billions of years ago.

Flat, open or closed

In 1929, Hubble proved our universe is expanding. In 1933, Zwicky noted that galaxies rotate faster than they should given how little mass we see in them. High schoolers learn Euclidean (flat) geometry. Parallel lines never touch and never diverge. Triangles have 180 degrees. A flat universe expanding slows but never stops. For an open universe, topologically like a saddle, parallel lines diverge and triangles have less than 180 degrees. For a closed universe, topologically like a sphere, lines circle back to make loops, which is why you always see the back of your head. Parallel lines cross. Triangles have more than 180 degrees. Over time, a closed universe stops expanding then collapses back. In 1998, we confirmed our universe is flat and the expansion is speeding up. We live in a finite, flat, closed, unbounded, expanding universe.

Pre-existing space

Does our universe expand into pre-existing space? No. **GR** tells us our universe creates space-time and matter-energy on its own as it expands. Today, the loop distance to the back of your head is bazillions of miles. At the big bang, loops were much shorter and tightly curved.

What causes the expansion?

Most folks believe the big bang was when all the energy present today suddenly poofed into existence. No. The big bang was not an explosion sending shrapnel into pre-existing space. It was the moment matter-energy split from space-time. This occurred in a larger universe outside of ours that continues to inflate. We are in a tiny bubble that stopped inflating and continues to expand slowly. The expansion rate is controlled by the **GR** equation with Einstein's 1917 cosmological constant moved to the other side and interpreted differently. $G_{\mu v}=kT_{\mu v}$ (1915) becomes $G_{\mu v}=k_1T_{\mu v}-k_2$ (today). Our universe did not expand because its contents were hot and under what we think of as pressure. The **GR** equation is built into our universe and determines whether space-time expands, contracts or stays static. The state of our universe provides the conditions the **GR** equation uses to decide.

Denser matter-energy curves space-time tighter

Did time early on flow at the rate it flows today? No. Tighter space-time curvature slowed the passage of time. The age of our universe is calculated in part by playing the expansion backwards to determine when it collapses into a tiny volume (but not into a point). Hawking determined there was no singularity where our universe had zero size.

Space-time between galaxies expands

Galaxies are roughly stationary. They have a small motion in seemingly random directions. If they were moving away from a center point (us for example), their mass would be increased. Galaxies further away would be more massive because they recede faster. However, none of them shows the **SR** mandated higher mass for this expansion speed.

Questions about GR

GR works for large distances. But does it hold up in the realm of the very small? Distance terms appear in the denominator of the **GR** equations. At tiny distances, dividing by nearly zero produces large (but not infinite) results. This does not appear to be a problem for **E&M** forces. These work fine at subatomic distances. But does the **GR** inverse squared distance falloff also work at or below atomic spacing? Is the added cosmologic term actually a constant? Because gravity is so weak, this is impossible to test with current technology. Also, **GR** is not a quantum theory. Most physicists believe it should be.

Quantum Mechanics

Quantum Mechanics **QM** studies the very small. Mix tiny, colored specks in a clear liquid. Under a microscope, they jiggle. Einstein's 1905 brownian motion paper explained this and proved atoms exist.

The periodic table

1 H																	2 He
3 Li	4 Be											5 B	6 C	7 N	8 O	9 F	10 Ne
11 Na	12 Mg											13 Al	14 Si	15 P	16 S	17 Cl	18 Ar
19 K	20 Ca	21 Sc	22 Ti	23 V	24 Cr	25 Mn	26 Fe	27 Co	28 Ni	29 Cu	30 Zn	31 Ga	32 Ge	33 As	34 Se	35 Br	36 Kr
37 Rb	38 Sr	39 Y	40 Zr	41 Nb	42 Mo	43 Tc	44 Ru	45 Rh	46 Pd	47 Ag	48 Cd	49 In	50 Sn	51 Sb	52 Te	53 I	54 Xe
55 Cs	56 Ba		72 Hf	73 Ta	74 W	75 Re	76 Os	77 Ir	78 Pt	79 Au	80 Hg	81 Tl	82 Pb	83 Bi	84 Po	85 At	86 Rn
87 Fr	88 Ra		104 Rf	105 Db	106 Sg	107 Bh	108 Hs	109 Mt	110 Ds	111 Rg	112 Cn	113 Nh	114 Fl	115 Mc	116 Lv	117 Ts	118 Og
			57 La	58 Ce	59 Pr	60 Nd	61 Pm	62 Sm	63 Eu	64 Gd	65 Tb	66 Dy	67 Ho	68 Er	69 Tm	70 Yb	71 Lu
			89 Ac	90 Th	91 Pa	92 U	93 Np	94 Pu	95 Am	96 Cm	97 Bk	98 Cf	99 Es	100 Fm	101 Md	102 No	103 Lr

Atoms are mostly empty space about 10^{-10} meters in diameter. They contain the atomic number of protons confined into a central nucleus (**1.7** to 12×10^{-15} meters) with electrons zipping around this in orbitals. Wall chart periodic tables show element names and other properties.

Element abundance

Our sun is a third generation star, so there are elements heavier than H, He, Li and Be. The next chart gives comparative abundance of elements in our solar system today. Some hydrogen (H, #1) has fused into helium (He, #2) in the interior of our star since it formed 5 billion years ago. A tiny part of the energy released by this nuclear fusion warms the Earth. Elements above #92 are made in reactors. These all decay.

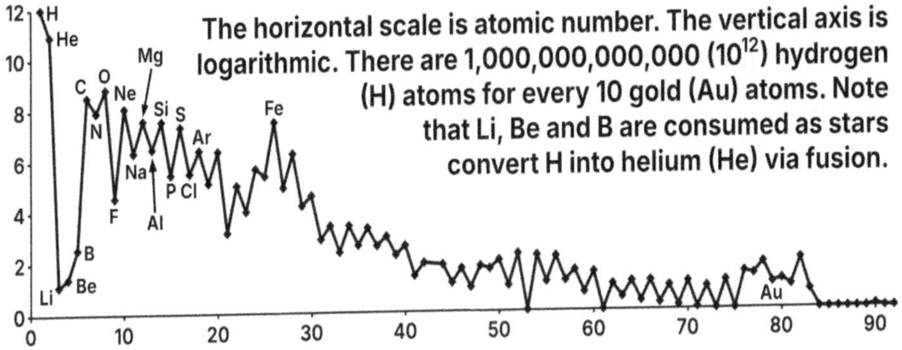

The horizontal scale is atomic number. The vertical axis is logarithmic. There are 1,000,000,000,000 (10^{12}) hydrogen (H) atoms for every 10 gold (Au) atoms. Note that Li, Be and B are consumed as stars convert H into helium (He) via fusion.

Neutrons

Protons are positively charged. Remember **E&M** fields? Like charges repel. To overcome this repulsion, neutrons (no charge) also reside in the nucleus. This increases the amount of strong nuclear force trapping protons and neutrons together inside the nucleus. Isotopes of an element have the same number of protons but a different number of neutrons. Protons plus neutrons is an element's atomic weight. Some isotopes are stable. Others decay over time into different elements.

Radioactivity

There are many nuclear reactions that change one element into another. Alpha decay occurs when a helium (He, #2) breaks off a larger nucleus. Helium is two protons and two neutrons. The original atom's atomic number drops by two. Its atomic weight drops by four. A piece of paper stops alpha particles. But if you ingest radioactive atoms, that paper is no longer between your cells and the decay energy. Beta decay is when a neutron splits into a proton, an electron and an antineutrino. The weak nuclear force governs this. The electron zooms from the nucleus. Atomic number increases by one. Gamma decay is a high energy photon emitted from a nucleus. This does not change atomic number or weight but does shed energy. These three common nuclear processes unshielded can cause cancer and other injury.

Half-life

The time for half of all atoms of a type in our universe to decay is half-life. This ranges from a fraction of a second to many years. If half-life is a day, after day 1, half the atoms have decayed. Day 2, 3/4 have changed, 1/4 are left. Day 3, 1/8 remain. Then 1/16, 1/32, 1/64 and so on.

Fission in nuclear bombs and power

If a U238 uranium, 92 protons and 146 neutrons, absorbs a neutron, it become U239, 92, 147. This beta decays to neptunium, 93, 146. Np239 beta decays to Pu239, plutonium, 94, 145. Atomic bombs use U235 or Pu239. When these capture a neutron, they split, spewing neutrons and energy. Critical mass is where each fission releases more than one neutron that splits another atom. Per fission, U235 yields 2.2 neutrons on average. Pu239 releases 6. For gun type bombs, 22 pounds of U235 or 6 pounds of Pu239 works. For implosion devices, even less. Power reactors use U235, Pu239 or, in the future, possibly thorium-232 (Th, #90).

Ions

There is an electron for each proton in atoms. Ions occur when electrons are stripped off or added. These are denoted with a superscripted **+**, **-**, **++**, **--**, **+**3, **-**3, and so on. Ions often combine to form crystals like sodium chloride, table salt (Na^+Cl^- or just NaCl).

Interpreting the periodic table

Electron orbitals come as **s**, **p**, **d** and **f**. The bottom two rows (8 & 9) of the periodic table tuck into the holes in rows 6 and 7 and, going left to right, fill electrons into **f** orbitals. Rows 5, 6, 7, 8 and 9 are rare, make colorful minerals and are used in metallurgy. Many folks have not heard of most of these. Column 18, rightmost, holds gases with their orbitals fully filled that do not bond with other atoms. After removing these rows and columns, the major players in chemistry remain.

1 H																
3 Li	4 Be											5 B	6 C	7 N	8 O	9 F
11 Na	12 Mg											13 Al	14 Si	15 P	16 S	17 Cl
19 K	20 Ca	21 Sc	22 Ti	23 V	24 Cr	25 Mn	26 Fe	27 Co	28 Ni	29 Cu	30 Zn	31 Ga	32 Ge	33 As	34 Se	35 Br

A row is an energy level. Electrons fill energy levels left to right (1s, 2s, 2p, 3s, 3p, 4s, 3d, 4p, 5s, 4d, 5p, 6s, 4f, 5d, 6p, 7s, 5f, 6d then 7p). The number of **s**, **p**, **d** and **f** orbitals (1, 3, 5, 10) sets the shape of the periodic table. Orbitals can contain two electrons, one with up spin, the other with down spin. Column one in the table, hydrogen (H, #1, 1s1) and column two beryllium (Be, #4, 1s2), plus the elements below these, fill the **s** orbital. Column 18 fills helium (He, #2, 1s2). Columns 13 to 18 fill three **p** orbitals. Columns 3 through 12 fill five **d** orbitals. For example, iron (Fe, #26) is 1s2, 2s2, 2p6, 3s2, 3p6, 4s2, 3d6. Left of carbon (C, #6) are metals like aluminum (Al, #13) that donate electrons. Metals with **d** orbitals, like copper (Cu, #29), conduct electricity better. Upper right elements pull electrons away from metals, ionizing them.

Covalent chemical bonding

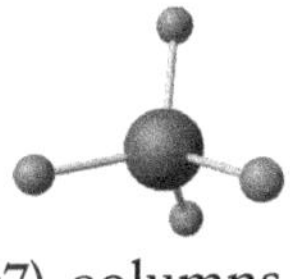

Atoms in carbon's (C, #6) column form four bonds. The ball and stick model is methane (CH4). Other things can replace the hydrogens. Boron's (B, #5) and nitrogen's (N, #7) columns form three bonds. Beryllium's (Be, #4) and oxygen's columns (O, #8) form two. Hydrogen's and fluorine's (F, #9) columns form one.

Particle in a box problems

$$i\hbar \frac{d}{dt}|\Psi(t)\rangle = \hat{H}|\Psi(t)\rangle$$

The Schrodinger equation gives us orbital geometry. Classic physics studies objects made from many atoms. **QM** looks at individual particles. The "box" limits a particle's location, for example, an electron in an orbital. Orbitals do not have sharp edges. As energy level increases, orbital size grows. Three **p** orbitals arrange themselves along **x, y, z** coordinates to keep the electron pairs as far apart as possible. Atoms tumble. So the **x, y, z** axes may orient in any direction. For bonding, one **s** and three **p** orbitals hybridize into four **sp** orbitals. Chemical bonds occur when **sp** orbitals overlap **s** or **sp** orbitals of other atoms. Hydrogen bonded to carbon is shown.

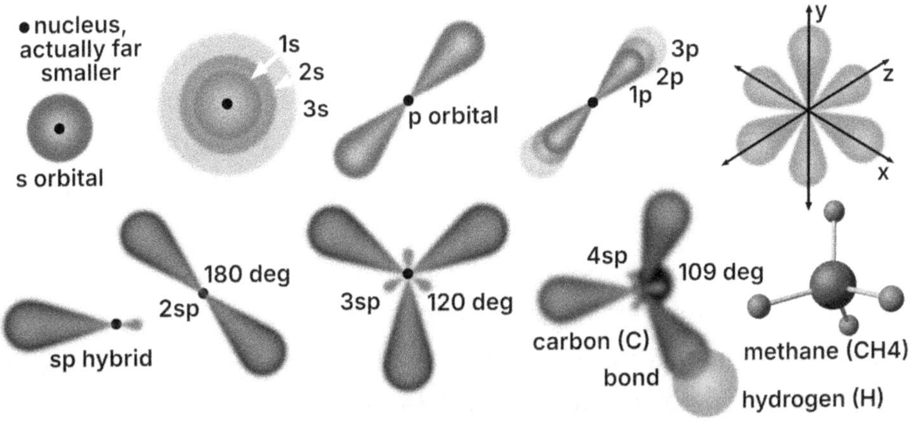

Carbon and silicon (Si, #14) bond tetrahedrally, as with CH4. Carbon makes flexible molecules. But pure carbon crystals (diamonds) are rigid, as are the nearly pure silicon crystals used to make transistors.

The molecules of life

Life, organic chemistry, uses predominantly carbon, hydrogen, nitrogen and oxygen. Most other elements are toxic. Some, like iron (Fe, #26), are found in molecules like hemoglobin. Other minerals, Na, K, P, Mg, Ca, Cr, Mn, Co, Cu, Zn, Se, Br and I (in row 5) are also used by life.

Photons

Photons are tiny pieces of light carrying energy and momentum. When a particle absorbs a photon, its energy and momentum transfers fully into the particle. The photon dies. To shed energy, particles emit new photons. If an electron in an energy level absorbs a photon, the electron could leap up one or more levels. It might emit one photon to carry the added energy away. Or, it could emit multiple photons to drop from level to level. If it sends multiple photons, each has less energy and the number of photons in our universe has increased. Same total energy but spread over more photons. Our universe has cooled and become more disorganized. Entropy has increased. This is how the second law of thermodynamics works at the **QM** level.

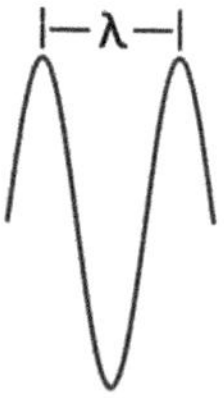

Heisenberg uncertainty principle

We cannot measure a particle's position and momentum exactly at the same time. To locate electrons, shine light on them. Light is many photons with energy $E=hf$ where $h=2\pi\hbar$, f is the frequency in cycles/second and $\hbar$ is 1.054×10^{-34} joule-seconds. To view a particle requires photons with wavelength λ smaller than what you wish see. But $\lambda=c/f$. High frequency, short wavelength photons carry lots of energy $E=hc/\lambda$. An electron absorbing a photon will change position and momentum. Δx is position uncertainty. Δp is momentum uncertainty. $\Delta x\Delta p\geq\hbar/2$. The uncertainty product $\Delta x\Delta p$ is always greater than $\hbar/2$. Δx and Δp can never be zero.

Why do electrons stay out of the nucleus?

Positively charged protons pull negatively charged electrons toward them. So why do electrons not fall into the nucleus and remain there?

If electrons lived in the nucleus, we would know both position and momentum simultaneously. The uncertainty principle forbids this.

Pauli exclusion principle

Two electrons cannot occupy the same energy state. However, electrons have a property called spin. An electron with up spin can share an orbital, or overlapping orbitals in a bond, with an electron having down spin. Without this exception, molecular bonds could not form.

Tunneling out of the box

Electric fields and the uncertainty and exclusion principles set the box boundary. You can never locate a particle precisely. But we can compute a probability field for where it might be. Adding the likelihood at every point sums to exactly 1.0. The particle must be somewhere. Given the fuzzy nature of orbitals, the probability field could extend outside the "box." Without adding energy, this allows a few particles to escape. We use this tunneling principle to fabricate useful devices.

Electrons never touch

When you sit, you float a small distance above the chair. Being closer to the surface than protons, electrons in your butt and the chair repel each other like tiny springs. You can never actually touch anything.

Transforming the uncertainty principle

The uncertainty principle was originally position versus momentum. As **QM** developed, uncertainty was recast in terms of energy versus time. The uncertainty in energy ΔE multiplied times the uncertainty in time Δt is again always greater than $\hbar/2$, yielding $\Delta E \Delta t \geq \hbar/2$.

Quantum leaps

In common parlance, a quantum leap is a big change. In physics, a quantum leap occurs when a single particle, an electron for example, absorbs a single photon and changes its energy level. This is a very tiny change. Be careful mixing common and scientific meanings.

Is quantum weirdness actually there?

Particles are quantum objects. They are an evolving quantum field, a thing unique to itself. Depending how we look, these might appear as tiny billiard balls or as waves smeared across space. A quantum field evolves until it exchanges energy with another quantum field. This transfer collapses the field, not some conscious entity observing. The universe evolved fine before physicists were here to watch. Is Schrodinger's cat alive or dead in its box with poison to be released at a random time? The cat's fate remains unknown until we open the box. In the double slit experiment, does a tiny billiard ball go through both slits? No. The quantum field does. Do not get trapped into the "is it this or is it that" false dichotomy. It could be either, both or something else. Reality is often a third option. Ditch the confusion. Quantum objects are quantum objects. Bazillions of quantum objects all doing their weird thing averages to the classic world we experience.

Entanglement

Physicists can entangle pairs of particles so they act as a single quantum object. If you measure a property of one, say spin, the other particle has opposite spin. If you invert the spin of one, the other particle's spin inverts instantaneously. This works even when the particles are separated by large distances. Is information being sent faster than light? **SR** prohibits this. But what if when particles are entangled, they connect via a **GR** wormhole? This solves the faster-than-light-speed communication concern. Particles un-entangle (de-cohere) when they interact with other particles, thus disconnecting the wormhole.

Determinism revisited

Many folks believe uncertainty is a measurement problem. No. It is a deep principle built into our universe. Normally never violated, the uncertainty principle can be overwhelmed as when stars collapse into black holes. Does uncertainty destroy determinism? We don't know. **QM** principles are differential equations. These are deterministic. We cannot measure perfectly. But our universe knows its state precisely. It is in that state. Bell's inequality (1964) ruled out local hidden variable theories. The 1982 Aspect experiment confirmed this. Did this kill determinism? Not sure. Do you have free will or only the illusion of free will? The debate continues. Is the **GR** wormhole work-around true? Don't know. If you are in 6th grade and choose to study diligently, maybe you will be the person who discovers this answer.

Particle Physics

QED is quantum electro dynamics, the study of electrons. **QCD** is quantum chromo dynamics, the study of quarks in the nucleus of atoms. Molecules, atoms and photons were the first quanta. Then we learned atoms had internal structure. In the 1950s and 60s as atom smashers became more powerful, many new particles were found. These were common during the big bang. But over the 13.8 billion years since, they all downgraded except the protons, neutrons and electrons making today's atoms. Electron neutrinos are also present in huge numbers. To explain the weird zoo of short lived particles, the standard model of particle physics emerged in the 1960s and 70s.

Leptons (6 types, all spin 1/2)

subtype	first	second	third
charged (-1)	electron	muon	tauon
neutral	electron neutrino	muon neutrino	tau neutrino

Leptons exist separately and are constantly in motion. The second and third columns are more massive and soon decay to the first types. Leptons respond to gravity and the weak nuclear force. Neutrinos rarely interact with matter, making them difficult to detect. Charged leptons (all minus one) also obey the electromagnetic force, the uncertainty principle and the exclusion principle.

Quarks (6 flavors, all also spin 1/2)

subtype	first	second	third
+2/3 charge	up (**U**)	charm	top
-1/3 charge	down (**D**)	strange	bottom

The second and third flavors also decay into the first. Quarks also move constantly. Three quarks make a particle. They are never found alone. Protons are one down quark and two ups (**DUU**). The two up +2/3 charges add to +4/3. The down -1/3 charge reduces +4/3 to +1, the proton charge. Neutrons are **DDU** and have no charge. The up +2/3 charge cancels the two down -1/3 charges. We never see +2/3 or -1/3 charges outside a particle. Like charges push two quarks apart. The opposite charged quark in the triplet pulls the other two toward it.

Force carriers

particle	what the force carrier does	spin
gluons (8)	mediates the strong nuclear force	1
W+	weak nuclear force	+1
W-	weak nuclear force	-1
Z	weak nuclear force	1
photon	carries electromagnetic force (**E=hf**)	1
higgs	confers mass to particles	0
graviton (?)	transfers space-time curvature	2

Quarks and leptons emit and absorb force carriers to transfer energy, momentum and other properties. All are quickly absorbed, except photons and gravitons. Photons and gravitons might be captured soon but can travel large distances before being absorbed. While gravitons are likely, their existence has not been confirmed. If they do exist, gravitons carry a minuscule quanta of space-time curvature.

Particles in the zoo

With 6 quarks at 3 to a particle, there are 110 unique, but not all the same quark, zoo particles (e.g., charm-down-strange, not up-up-up).

Spin

Leptons and quarks have spin 1/2. This carries angular momentum equal to 1/2 ħ. Since total angular momentum in our universe is zero, total spin is also zero. Particles made from quarks also have spin 1/2. Spin is labeled up or down for electrons. Spin is actually polarized right circular or left circular around a particle's axis, which is like an arrow. An arrow rotated perpendicular to its length 360 degrees looks the same as before it was rotated. However, if you rotate a 1/2 spin particle 360 degrees, its arrow reverses direction. You must rotate 720 degrees to make it look the same, weird but true. When leptons and quarks absorb or emit force carriers, spin direction could reverse.

String theory

String theory claims the standard model particles are tiny vibrating strings open ended or closed into loops. This gained adherents when an early version popped the graviton out of its equations. String theory looks promising. But none of its predictions have been confirmed.

Color, charge and spin carried by gluons

Quarks have a property called color: red, green or blue. Particles have a red, green and blue quark, making the particle white. This color analogy is not perfect. There are 8, not 9, gluons, no {blue-antiblue}. Gluons pair a color with an anticolor. When a green quark absorbs a {red-antigreen} gluon, it turns red. Quarks change color charge and spin continuously by emitting and absorbing gluons. For spin, **-1/2+1→1/2** and **1/2+1→-1/2**. Color and charge forces keep the three quarks inside particles jostling roughly $\mathbf{10^{-15}}$ meters apart.

Mass

particle	**mass**
electron	0.511
muon	1,056
tauon	1,777
e neutrino	$< 5\times10^{-8}$
m neutrino	**?**
t neutrino	**?**
up (**U**)	2.16
down (**D**)	4.70
charm	1,273
strange	93.5

particle	**mass**
top	172,570
bottom	4,183
gluon	0
W (+ & -)	80,369
Z	91,188
photon	0
higgs	125,200
graviton (?)	0
proton	938
neutron	940

Mass is given as millions of electron-volts divided by light speed squared. Existing theory thus far cannot calculate this. Mass must be measured. The higgs field imparts mass to particles. Neutrinos rarely interact with matter, so their mass is difficult to measure. Particles with zero mass travel at the speed of light until they are absorbed. These carry energy and momentum but **SR** forbids mass. At light speed, if they had mass, this would become infinite, which is not possible.

Antiparticles

In 1928, Dirac predicted an antiparticle for every particle. These have the same mass as their particle, the opposite charge and the opposite spin. When a particle meets its antiparticle, they annihilate. Their energy transfers into a photon. Note that a particle traveling forward in time is the same as its antiparticle traveling backward in time and vice-versa. Equal numbers of particles and antiparticles were likely produced by the big bang. So why is matter plentiful today while antimatter exists only when it is made during high energy interactions in particle accelerators and in the interior of stars? Evidence indicates the weak force is not fully symmetric.

Proton and neutron mass

A neutron is two down quarks and one up, **DDU**. Adding the **D** and **U** masses in the table gives (**4.70+4.70+2.16=**) 11.56. Dividing this by the electron mass yields (**11.56/0.511=**) 22.6. But the mass of protons and neutrons has been measured to be 1,836 and 1,839 times the electron mass. Most of the proton and neutron mass is caused by temporary virtual particle pairs inside protons and neutrons. How's that for weird?

Proton and neutron uncertainty

As with electrons, momentum and position cannot simultaneously be known. Because they are more massive than electrons, staying within a tiny nucleus does not violate the uncertainty principle $\Delta x \Delta p \geq \hbar/2$ limit. Momentum **p=mv** depends on mass **m.** Larger mass shrinks the volume into which protons and neutrons can be legally confined.

Weak nuclear force

The **WNF** causes quarks and leptons to change type. The two main interactions are **#1,** a neutron decays into a proton, electron and an antineutrino (beta decay) and **#2**, a high energy proton merges with an electron and converts into a neutron and neutrino. In either case, energy is shed, keeping thermodynamics happy. Energy and momentum always balance. Momentum is **p=mv.** The energy in motion is $mv^2/2$. Students use these facts and particle masses to calculate velocities. To complicate things, if particles move near the speed of light, the **SR** increased mass and half-life must be used.

Making neutrons

A neutron is **DDU**. A proton is **UUD**. Reaction **#1** changes a **D** to **U**. Reaction **#2** changes a **U** to **D**. Across our entire universe, every 614 seconds, half the neutrons not in a nucleus decay via reaction **#1**. When bound in a nucleus, neutron half-life extends greatly. Note that a particle's half-life increases when it moves fast, as **SR** dictates. Also, a **U** is lower mass than a **D** making the **D** to **U** change thermodynamically favored. To go the other way, **U** to **D**, the proton must be energetic (hot). Inside stars where it is very hot, four hydrogens (1 proton each) fuse into a helium (2 protons and 2 neutrons). For this, we need more neutrons. Reaction **#2** produces these. Roughly 100 trillion, $\mathbf{10^{14}}$, neutrinos pass through your body every second without you noticing.

Feynman diagrams

These visualize particle interactions. Time passes moving bottom to top. Horizontal is distance between particles. In the first example, two electrons will collide. **e1** sends a photon to **e2** changing their paths. Thought of as billiard balls, leptons and quarks never touch. But their quantum fields, which spread out a bit and are slightly fuzzy, overlap. In the second, a neutron **n** emits a $\mathbf{W^+}$ force carrier and changes to a proton **p**. The $\mathbf{W^+}$ decays into an electron **e** and antineutrino $\bar{y}$. Both zoom out of the nucleus. Again, this is beta decay.

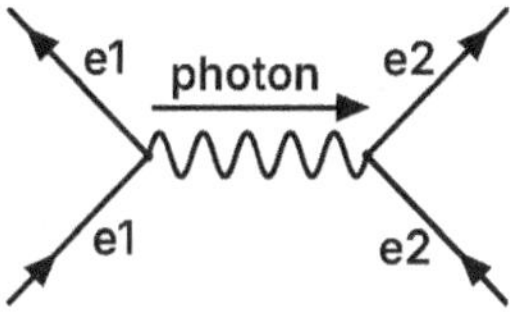

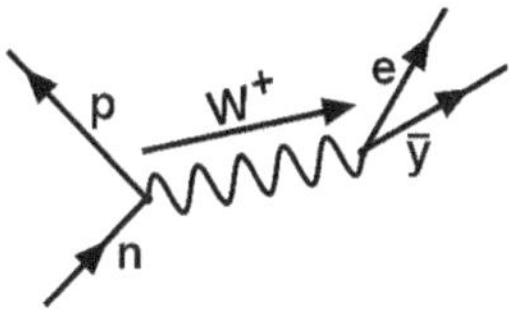

Virtual particles

While thermodynamics demands energy interactions between particles balance, $\mathbf{\Delta E \Delta t \geq \hbar/2}$ permits a loophole. A particle and antiparticle pair can pop into existence so long as they annihilate before an uncertainty principle time violation occurs. Our universe can borrow energy if it pays it back quickly. Virtual particle pairs pop in and out of existence regularly then quickly recombine. Their collective mass in empty space (nothing) is dark energy, which causes the **GR** extra term that provides an antigravity force. Otherwise, gravity would always be attractive. Our universe expands because empty space (nothing) weighs something. How's that on the weirdness scale?

Symmetry

Quanta

Light, atoms, neutrons, protons, leptons, quarks and force carriers plus properties like charge and spin come in tiny packets. Even gravity, time and space may be quantized. Is there a smallest unit of time? Do we leap from moment to moment? Is there a smallest unit of space?

Conservation laws

Conservation means at the start and end of interactions, the same amount of the conserved thing is present, though its distribution across particles has changed. Conservation laws include momentum, angular momentum (spin), energy and charge. All momentum, angular momentum, energy and charge summed across the entire universe is zero.

Symmetry

An example of **QM** symmetry was rotating a spin 1/2 particle 720 degrees to return it to its initial appearance. Unmarked spheres rotated always look the same. A square rotated 90 degrees looks the same. In every direction, water looks the same. Ice crystals are hexagonal from one direction and flat from another. Symmetry was broken by freezing water to ice, the same molecules look different. Our universe is also symmetric. Through a telescope, galaxy distribution in every direction is similar. Nearby, our milky way galaxy appears as a band of stars. Seen near, different. Seen far away, the same. This is symmetry in physics.

Phases of matter

Classic physics identified four phases of matter, solid, liquid, gas and plasma. Most people understand the first three. A plasma is when temperature is high enough for electrons to be stripped from around the nucleus. This gives a hot soup of electrons, naked nuclei and photons. If an electron emits a photon, it may be captured into an atom. The electron then absorbs a photon and may again be free. Stars are plasmas. Their “surface” occurs when temperature drops as energy is radiated away. Electrons can then be captured by nuclei. The plasma becomes a gas and becomes transparent to light. Elements show weird properties like superfluidity and superconductivity near absolute zero. This could be called a fifth phase. To change from solid to liquid, liquid to gas or gas to plasma, energy (latent heat) must be added to break bonds. This is why you can boil water in a paper cup over a campfire. The surface of the sun is not solid. It is plasma transitioning to a gas having a much lower density than you. Standing on it, you would sink.

Four forces

We know about gravity, electromagnetism, the weak nuclear force and the strong nuclear force. Could there be a fifth force? Possibly, but there is no solid evidence thus far to support this. Could a fifth force be involved with dark matter? Maybe. We don't know.

Symmetry breaking

In 1918, Noether wrote a paper equating conservation laws with symmetry breaking. When a symmetry in our universe breaks, a conservation law is born. Time translation symmetry broken gives energy conservation, the first law of thermodynamics. Space translation symmetry broken gives conservation of momentum, Newton's third law. Rotational symmetry broken gives angular momentum and spin conservation. Gauge symmetry broken gives charge conservation. Physicists rely on conservation laws to solve problems every day.

Weak nuclear force asymmetry

To answer why matter remains today and was not annihilated completely when it mixed with antimatter following the big bang, the **WNF** appears not to treat matter and antimatter equally. Without this imbalance (an asymmetry), we would not be here. Is the missing blue-antiblue gluon an asymmetry of the strong nuclear force?

The arrow of time

We experience time flowing in one direction. Total entropy always increases in accord with thermodynamics law **#2**. But the equations of physics are time symmetric. Particles and antiparticles can and do travel forward or backward in time. There is nothing to prohibit a broken cup from jumping up and re-assembling on a table. The same is true for the gas in our earlier box after it expanded to fill both halves. Let's say there are 100 particles total bouncing around. The most likely case is half on each side. There is one state with all 100 in the left, 100 states with 99 in the left half and one in the right, (**100×99=**) 9,900 states with two in the right, (**100×99×98**=) 970,200 states with three on the right and so on. Now do this for more than $\mathbf{10^{20}}$ particles. Every state is equally possible. There is one state with zero particles in the right and bazillions of states where particles are more evenly balanced. Equal amounts on each side partly defines thermal equilibrium. This is one reason why, at human scale, we experience time flowing only in one direction and broken cups do not spontaneously reassemble. For **QM**, the rules are more permissive. Add more weirdness.

Force de-unification

At the big bang, space-time split from matter-energy. At that instant, energy was tightly packed. Particles, as we know them, did not have room to exist. Things were very hot. Excluding gravity, the other three forces in this high energy environment acted as a single unified force. Because the **GR** field equation dictated this expansion, our universe spread out. Its size increased. Things cooled. The same energy spread out over more volume. Soon the strong nuclear force and the electro-weak force separated. Things cooled further. Then the electro-weak force split into the weak nuclear force and the electromagnetic force. Each of these bifurcations was a phase change, a symmetry breaking, like steam condensing into water then freezing into ice. Interesting physics happens at these transitions.

CPT symmetry

C stands for conjugation, changing particles into antiparticles. **P** is parity, where spacial coordinates are flipped, creating a mirror image. **T** is time reversal, exchanging future to past and vice versa. The laws of physics remain unchanged when all three transformations **CPT** are applied together. For our universe to maintain **CPT** symmetry, asymmetries in **C**, **P** or **T** individually are compensated for by asymmetries in the other two. This relationship, proven in the 1950s, is a foundation stone for the standard model and is actively researched.

The Big Bang

Most folks believe the big bang was when a huge amount of energy suddenly poofed into existence. Wrong. Our universe has zero net energy. Was the big bang an explosion sending shrapnel into pre-existing space as most science programs show? No. Our universe makes its space-time as it expands. Could you watch the big bang from outside? No, our universe is closed. Nothing, not even light, gets in or out. Our universe is a tiny bubble that stopped inflating in a far larger cosmos. The parent universe outside our universe continues to inflate.

Evidence for the big bang

Before **GR**, folks believed our universe was static and eternal. Reading Einstein's 1916 **GR** paper, a few physicists predicted it is expanding. Then in 1929, Hubble used the doppler effect to measure the motion of galaxies. Our universe is in fact expanding. Reverse time. Play the video backwards. Everything collapses back into a small volume. Time had a beginning. Expansion is the first proof for the big bang.

More powerful telescopes

Before Mount Wilson's reflecting telescope, we could not photograph galaxies (fuzzy nebula) with resolution sufficient to see individual stars. Hubble published photos of galaxies but his most notable work was with the spectra of light coming from galaxies outside our own.

Doppler effect

Sound from vehicles, trains in 1842, is higher pitched as it comes toward you and lower pitched as it recedes. Similarly, the spectrum of light emitted from objects approaching is blue shifted while light from objects moving away is red shifted.

Light spectra

All warm bodies emit radiation from low frequency, low energy radio waves to microwaves, to infrared, to visible light (red, yellow, green, cyan, blue, magenta), to ultraviolet, to high frequency, high energy x-

and cosmic rays. Light spreads into its spectrum when passed through a prism. Given the emitting temperature, we can compute the number of photons at each frequency. Magnifying the spectrum, we find bright and dark lines, the quantum fingerprint of an atom's allowed electron energy levels. This tells us which atoms emitted the light.

Space-time itself is expanding

Hubble compared dark lines in light spectra. The more distant the galaxy, the more these lines shifted red. This red shift is the same looking in every direction. No, we are not the center from which other galaxies are flying away. From every other vantage point, the red shift is the same. None of the galaxies show the **SR** required mass increase due to acceleration to their observed speed. If they were that massive, their gravity would be stronger. In reality, galaxies are roughly stationary. The space-time between the galaxies is what is expanding.

What happened during the big bang?

The big bang occurred earlier in a larger universe that contains ours. Space-time and matter-energy split in that larger universe and remain split in ours. The parent universe continues to inflate enormously fast. The moment our bubble's space-time stopped inflating, a huge amount of energy was packed into a tiny volume. This signaled our universe's built in **GR** to expand at a moderate rate. Back then, energy density was far too high for particles as we know them.

Planck values

Physics we know breaks down before the planck time, $\mathbf{5.391\times10^{-44}}$ seconds, the square root of $\mathbf{hG/2\pi c^5}$, and below the planck length, $\mathbf{1.616\times10^{-35}}$ meters, how far light travels in the planck time. The plank temperature is $\mathbf{1.427\times10^{32}}$ degrees Kelvin. At this, three forces remain unified into a single force. Gravity likely de-unified from the other three forces earlier in our parent universe. Particles heavier than the planck mass, $\mathbf{2.176\times10^{-8}}$ kilograms, collapse into black holes.

Three more forces de-unify

Things in our bubble expanded and cooled until the strong nuclear force separated from the unified electro-weak force. Expansion continued. The electro-weak force then split into the weak nuclear and electromagnetic forces. Now there were four forces. Enough space soon became available and temperature was low enough for particles made from third column quarks and leptons. Particles and antiparticles formed in equal or nearly equal amounts. These then annihilated, leaving roughly one matter particle for every billion, **10^9**, photons.

More expansion

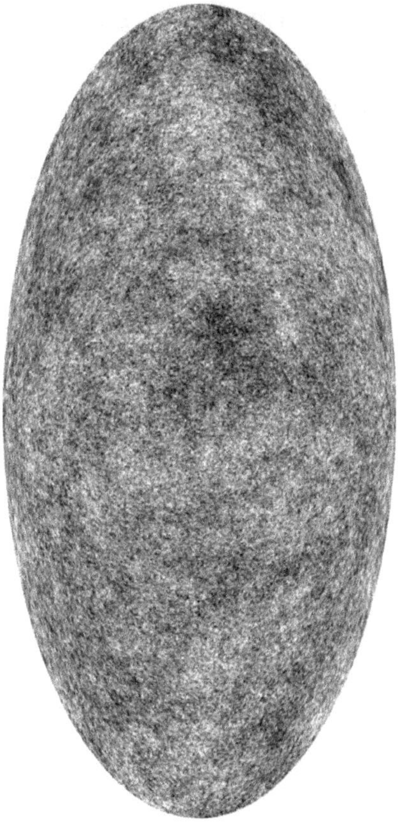

Our universe continued to expand. Zoo particles from second and third column quarks formed. These second and third column types decayed into first column varieties to form neutrons **DDU** and protons **DUU**. We had a hot plasma soup of zoo particles, protons, neutrons, electrons, neutrinos, dark matter and force carriers zipping about. Nucleosynthesis made some He, less Li, Be and B and very little else. When the universe was a few minutes old, fusion stopped. The temperature was now below that found in the interior of stars. Things continued to expand for another 380,000 years as temperature fell to roughly 3,000 degrees Kelvin where electrons can drop into orbitals forming hydrogen atoms. Our universe became transparent to light. The baby **CMBR** (cosmic microwave background radiation) photo shows our universe at that time. The gray scale indicates regions of greater and lesser temperature and density.

Entropy

Entropy is second law speak for disorder. As time passes, things go from orderly to disorderly. Our universe post big bang had 75 percent hydrogen by mass, 25 percent helium, photons and dark matter. This is very orderly. Our cosmos is now much larger with more kinds of stuff spread around. This is less orderly. Energy density is also much lower. Use nuts, bolts, etc. in jars as an analogy. This is orderly. On a workbench, smash the jars, mix this stuff up and spread it out. Our universe is now less orderly, meaning higher thermodynamic entropy.

What else was present?

Dark matter helped galaxies form. Mini black holes and magnetic monopoles might also have existed. However, most mini black holes would have evaporated by now. Magnetic monopoles, north and south, may have annihilated each other. If these existed, Maxwell's equations might have been fully symmetric back then.

Inflation

In 1980, Guth proposed inflation. At the big bang, the universe expanded slowly, remaining causally connected for a time. This means tiny enough for energy to slosh around at light speed many times as things came to nearly the same temperature. Then a phase change occurred, possibly when the strong nuclear force split from the electro-weak force, a symmetry breaking. By analogy, when water cools, extra energy, latent heat, must be removed before it can freeze. With no impurities or vibrations, temperature can drop below 32 degrees F (0C) before ice crystalizes. Tapping on the container freezes it. This phase change latent energy could cause antigravity. Nothing can travel faster than light. But there is no limitation on how fast space-time can expand. In a fraction of a second, the universe inflated to well above 10^{25} times its original diameter. Inflation then stopped with the horizon, flatness, **CMBR** temp, galaxy formation and monopole problems solved. Hawking saw a flaw in Guth's method for ending inflation. But if our universe was a bubble within a larger inflating parent universe that stopped inflating, this glitch is fixed. Our bubble came from a spot in our parent universe already near to but not at thermal equilibrium.

Galaxy distance

Measuring distance to galaxies is difficult. We now have better telescopes and standard candles, stars shining at predictable brightness. Type-1A supernovas are extremely luminous for a month. Telescopes can view 100,000 galaxies at a time. Galaxies contain hundreds of billions of stars. That's 10^{15} stars, enough to spot a few type-1A supernovas each night. The brightness (red shifted) measured gives distance.

From this data, we know the expansion rate today is increasing and no longer slowing from the effect of gravity, as previously believed.

Einstein's cosmological constant

Einstein added a factor to **GR** to make our cosmos static. But this term is unstable, like balancing a pencil on your finger. It must constantly be tweaked or our universe will expand or contract. Because we know the expansion is speeding up, a constant, or something like it, must be included in **GR** to produce an antigravity force that accumulates over large distances. Within our galaxy, this effect is miniscule.

Casimir effect

How do we know nothing is unstable? Put two tiny metal plates near each other in a well shielded vacuum. The plates move to each other. Why? Because nothing pops up pairs of virtual particles constantly that exist for less than the time allowed by uncertainty. Virtual particles "hit" the plates more outside than between producing a net inward force.

CMBR blackbody spectrum

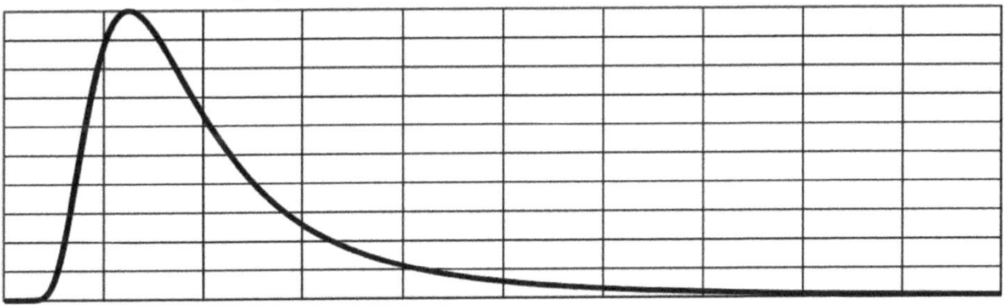

In 1879, the energy radiated from warm bodies was determined to be $\mathbf{E=A\sigma T^4}$, where **A** is the surface area, $\sigma=5.67\times10^{-8}$ watts/m^{2}k^4 and **T** is temperature in degrees Kelvin (k). In 1900, Planck deduced a formula for the spectrum of light from a blackbody radiator. In 1964, the cosmic microwave background radiation was discovered. Our universe's baby photo shows a 2.7 degree kelvin blackbody. The **CMBR** temperature variance (the grayscale) is roughly 1/10,000 of a degree. This light started toward us when our universe cooled to about 3,000 degrees and became transparent. At 3,000 degrees, the wavelength range for the horizontal axis of the chart is from **0** to $\mathbf{8.2\times10^{-6}}$ meters. For the stretched **CMBR** light, the range is **0** to $\mathbf{8.6\times10^{-3}}$ meters. Both curves are identical. However, the energy radiated was $\mathbf{(3000/2.7)^4 = 1.5\times10^{12}}$

greater. Hotter bodies radiate much more energy. Our universe has expanded to (**$8.6\times10^{-3}/8.2\times10^{-6}=$**) 1,049 times its **CMBR** size. This stretched the wavelengths from visible light into the microwave range. The **CMBR** is the second major big bang proof.

Photon energy stretched out

Photons have energy **$E=hf$**. The above chart is for wavelength **λ**. But **$\lambda=c/f$**. So **$E=hc/\lambda$**. The **CMBR** blackbody spectrum photons were emitted as visible light and have stretched. The wavelengths humans can see range from **380×10^{-9}** to **750×10^{-9}** meters. Higher frequency, shorter wavelength light carries more energy per photon. So where did the energy in the **CMBR** photons go? It went into climbing out of a gravity well. This slightly flattens space-time. Light leaving Earth also red shifts a bit as it works to escape Earth's gravity. This was predicted by **GR** and has been verified, yet another **GR** validation.

Calculating post big bang element abundance

Using the standard model of particle physics, we can calculate the H to He to Li to Be to B ratios. This is the third major big bang proof. No other model ever proposed passes these three (and other) tests. Science can never prove anything absolutely. But as confirming evidence accumulates, we gain confidence that a theory is correct.

What happened before the big bang?

Nothing. But **QM** tells us nothing is unstable. It appears nothing popped out a parent universe with zero net energy from which our bubble universe stopped inflating. Given zero net energy, there is no violation of conservation of energy (thermodynamics' first law) or of the uncertainty principle. A zero-energy universe can exist forever.

Radio and other telescopes

Because our universe is expanding, light, as humans see it, from distant objects has stretched toward the redder end of the electromagnetic spectrum. Radio telescopes collect longer wavelength light. These are physically larger than optical telescopes. Most radio telescopes are aim-able dishes. You can link these together to create a larger radio telescope. We also build x- and cosmic ray detectors and fly these in space to avoid Earth's electromagnetic noise pollution.

The horizon

We cannot see our entire universe. And we can never look outside our bubble. The topology of space-time forbids this. The light we see in the **CMBR** has been coming at us for 13.8 billion years. A light-year is how far light travels in a year, $\mathbf{5.88 \times 10^{12}}$ miles. Because our universe has expanded, the sphere around us where the **CMBR** light originated has grown to $\mathbf{9.3 \times 10^{10}}$ light-years in diameter. When objects recede, their light reddens. When that velocity exceeds light speed, the frequency drops to zero. We can never see past this horizon. As our universe expands further, we will view less of it. Prior to the **CMBR** light release, our universe plasma was opaque. To see beyond this **CMBR** barrier we could use gravity waves. We have detected them. But forming an image using gravity waves is unlikely during my lifetime.

The dark age

When our universe cooled to where the plasma collapsed into atoms, there was a bright white glow that dimmed and reddened as our universe expanded. Darkness fell until local patches of denser hydrogen and helium collapsed inward due to gravity. The lights turned back on as stars began fusing hydrogen into helium.

Some temperature variation needed

Had our universe started at full thermal equilibrium, all turbulence damped out, stars and galaxies would not form from gravity collapse of denser regions. That 1/10,000 degree variance made galaxies happen.

Time at the big bang

Big bang space-time was far more tightly curved. So time passed more slowly, as it does when falling into black holes or traveling near light speed. Living in the Earth's mild gravity slows time slightly.

Post CMBR

Post **CMBR**, we had an expanding cloud of hydrogen, helium, some Li, Be and B, photons, neutrinos and dark matter. Something like dark matter must exist for our universe's **GR** solution to match observations. Because the larger cosmos that spawned our bubble universe had not come fully to thermal equilibrium, turbulence remained. We had higher and lower density and temperature regions. The denser patches collapsed into stars, large groups of stars (galaxies) and clusters of galaxies. The clusters appear to be distributed along the surface of what appears to be bubbles making a cosmic foam. These bubble interiors are largely devoid of galaxies. This fossil hints at earlier dynamics.

The expansion of our universe is speeding up

During the 1990s, to balance the **GR** equation to match observations, a few physicists predicted, rather than gravity slowing the expansion down since the big bang and into the future as expected, the expansion was actually speeding up. This has now been confirmed.

Dark matter and dark energy

With the added **GR** antigravity term, normal matter-energy we can inventory makes up 5 percent of our universe. Dark matter, the stuff needed to make galaxies rotate correctly is 27 percent. Dark energy, the stuff causing the antigravity force to speed up, is 68 percent. Dark energy is very likely virtual particles popping in and out of existence. Because these short-lived matter-antimatter pairs have mass, empty space-time (nothing) weighs something. This weight is accelerating the expansion. The average density of normal matter in our universe is roughly five protons per cubic meter. Most of our universe is very empty. Normal matter energy density is $\mathbf{0.52\times10^{-33}}$ g/m^3. Dark matter energy density is $\mathbf{2.8\times10^{-33}}$. Dark energy density is $\mathbf{7\times10^{-33}}$.

How big is our universe?

There are $\mathbf{10^{86}}$ particles in the observable part of our universe. But our universe extends beyond the **CMBR** horizon. The observable part up to the **CMBR** is a sphere $\mathbf{5.5\times10^{23}}$ miles in diameter embedded within our universe, which is far larger. This is then embedded in an even larger parent universe still inflating outside our universe. Could there be other universes in the larger inflating parent universe? Yes, very likely. We have gone from small and recently created to enormous, but not infinite, and to 13.8 billion years old.

"Nothing" is unstable

QM and particle physics have demonstrated "nothing" pops out particles. One time long ago when this happened, space-time also split from matter-energy and then inflated. This appears to be how we got here.

Stars and Galaxies

Star generations

First generation stars have H (1), He (2) and tiny amounts of Li (3), Be (4) and B (5). Heavier elements are forged as stars die. Later generation stars contain increasing amounts of heavy elements.

High energy chaos early on

When our cosmos was small, large stars were more likely. The bigger the star, the more gravity presses down and the faster hydrogen fuses to counter this. Early on, our universe had many huge stars with short lifetimes that spewed heavy elements. Also, stars crashing into each other formed huge black holes now found at the centers of galaxies.

Main sequence

Stars balance the force of gravity against the heat released by fusion. Over time as the hydrogen is consumed, the star's diameter gradually shrinks and the star becomes brighter. All garden variety stars do this.

Death of stars

When stars run out of fuel, they collapse fusing heavier elements, then expand into red giants. In our solar system, this will end life on the Earth. Stars like our sun, or smaller, then re-collapse to finish as white dwarves roughly 7,000 miles in diameter. These slowly cool. Electron exclusion sets the radius. Somewhat larger stars overwhelm electron exclusion. In the remaining core after these explode upon re-collapse, electrons are crushed into the nucleus converting protons into neutrons to form neutron stars roughly 12 miles in diameter. Quark exclusion sets the radius. Even bigger stars re-collapsing explode with their core being crushed into a black hole. The density of Earth is 5,500 kg per cubic meter, of our sun, 160,000, of a white dwarf, 10^9 and of a neutron star, 6×10^{17}. For black holes, density depends on its mass. Larger stars leave a neutron star or black hole in their dying explosion. This blast terminates nearby life to a much greater radius.

Dying stars spew a large part of their material. Over time, this gas and dust re-collapses into new stars and solar systems.

Conservation of angular momentum

Galaxies and stars orbiting and rotating conserve angular momentum, including the post big bang turbulence. Having a much smaller diameter, white dwarves, neutron stars and black holes spin far faster than the stars that died to form them.

Mass defect

Helium mass is 4.00260 atomic mass units. The mass of 2 protons, 2 neutrons and 2 electrons is 4.03298 amu. The 0.003038 difference is the mass defect. To break helium apart requires this much energy. It is also the energy released when 2 hydrogens and 2 neutrons fuse into helium. So, only (**100×0.003038/4.03298=**) 0.075 percent of the mass can be converted to heat by fusion. Remember $E=mc^2$? Fortunately, c^2 is a very big number and stars are huge. Our sun has shined five billion years and has hydrogen fuel for another five billion. Atoms up to iron release energy via fusion. This is why iron (Fe, #26) peaks on the page 19 chart. Heavier atoms shed energy via radioactive decay and fission.

No 5 or 8

No stable isotopes with atomic weight 5 or 8 exist. Helium (2 protons, 2 neutrons) is very stable. Atomic weight 8 atoms soon split into two heliums. Atomic weight 5 atoms eject a proton or neutron. So, how do heavier elements form? When two heliums fuse, beryllium-8 survives long enough for a third helium to merge, forming carbon-12. Heavier elements are made by adding heliums to produce oxygen-16, neon-20, magnesium-22 and so on. Ditto for atomic weight 5 atoms.

Failed stars

Large gas planets are stars that failed to accumulate enough mass to fuse hydrogen into helium. Jupiter, Saturn, Uranus and Neptune, all gas giants, are failed stars. We have found many failed stars orbiting nearby stars. Also, binary star systems frequently form. Our nearest solar system neighbor is three stars, Proxima Centauri, Alpha Centauri A and Alpha Centauri B all circling each other. Gravity and momentum conservation produces a wide variety of solar systems.

Going Local - Planets

When our sun, a third generation star, formed, it received 99.9 percent of the mass from the cloud of dust and gas that collapsed under gravity. Our planet accreted from a tiny fraction of the remaining 0.1 percent. Accretion means the Earth collided with the material along its orbital path around the sun, sweeping its path clean.

Big rocky planets inside

Planets are big enough for gravity to make them spherical. They also orbit in the plane of the solar system. Pluto was captured from the Oort cloud beyond our solar system. Its orbital plane tilts 17 degrees. If a star or stars get most of the hydrogen that collapsed, they also got most of the heavy stuff. While stars are turbulent, heavier elements do tend to migrate inward. This is also true for large gas planets. At the center of Jupiter, there is a sizable rocky planet under its thick atmosphere.

Goldilocks zone

We occupy a planet where energy from our star warms us enough for liquid water. If the Earth was too close or too far from the sun, we could not live here. The main sequence is a concern. Stars brighten as they age. When our sun formed, it was 40 percent dimmer than today. The goldilocks zone slowly moves outward. With no greenhouse effect, our surface temperature would be 55 degrees Fahrenheit (31C) colder. Our oceans would be frozen over. This is a factor in the global warming debate very few people discuss. Fortunately, climate change is not occurring as fast as anti-fossil-fuel activists forecast. But it is happening, which the deniers deny. The truth is usually in the middle.

Earth and planets like it

We live on a planet big enough for gravity to hold a thin atmosphere, 78 percent nitrogen N2, 21 percent oxygen O2, 1 percent argon Ar plus tiny amounts of other gases like carbon dioxide CO2, water vapor H2O and methane CH4 plus pollution. O2, N2 and Ar are transparent to most light from the sun. N_2 and O_2 scatter blue light but do not absorb it, making our sky blue. During scattering, if the photons emitted have the same energy as those absorbed, the interaction is 100 percent efficient. No energy is converted to low grade heat. If N_2 and O_2 absorbed solar energy, our air would be too hot for life. Venus suffered a runaway greenhouse effect. Our gravity is too weak to hold hydrogen, helium or neon. These were abundant when the Earth formed but have mostly trickled away. Some hydrogen remains as water H_2O. The trace gases regulate our surface temperature. Our air and magnetic field protect us from harmful radiation. The Earth has cooled. So earthquakes are mild. Our weather is also tame. For us, the Earth is the best of all planets.

Climate

Over our 4.5 billion year history, our environment has evolved. Early on, the sun was dimmer. A larger greenhouse effect was needed for liquid water to exist. Huge amounts of methane CH_4 in the primitive air provided this. Then life happened. As the sun brightened, we did not overheat because life buried some carbon C from the air. Then chlorophyll and plants emerged that release oxygen, O_2. This reduced the greenhouse effect by turning CH_4 into carbon dioxide CO_2 and water H_2O. There were eight snowball earths. As plants died back, O_2 output declined, so we unfroze. As the sun warmed further, a CO_2 only greenhouse effect was sufficient. Nearly all carbon has now been removed from the air to keep us cool. Plants eat CO_2 and are starting to starve.

Other solar systems

The Earth and our sister planets are likely typical throughout our universe. Looking out also looks back in time. Young galaxies were violent places not suitable for life. Looking nearer to us and not so far back in time, things settle out to where later generation stars can support life. The distant galaxies forming long ago are now like what we see nearby. Life is likely abundant throughout our universe. But sentient life capable of technology would be far less plentiful.

How did our moon get here?

As our Earth accreted, its surface was molten from the energy of gas, dust and objects impacting. Due to gravity interactions, not all things orbited uniformly in the solar disk. We were struck by a large object, shattering our planet. Most of the material fell back to re-form Earth. But some remained in orbit, accreting into our moon. Is a moon essential for life? No. But tides and currents the moon's gravity causes bring minerals from the ocean bottom to the surface making them available for life. The presence of chlorophyll and life in ocean surface waters correlates with ocean currents and land mass edges. Note that this impact tilted Earth's axis by 23.5 degrees, causing the seasons.

Other solar systems

We have found many planets around nearby stars. So it appears most stars have planets. Also, there are lots of moons around the planets in our system, indicating moons likely orbit planets around other stars. Most of the planets discovered elsewhere are large. Big planets are easier to detect. But we have found smaller rocky exoplanets with some orbiting their star(s) in the goldilocks zone. The distribution of large and small planets around stars varies. In addition to planets, our sun has rings like Saturn. This is our asteroid belt. It is there because Jupiter's gravity interferes with asteroids forming into a planet. Saturn's rings are present because gravity from its moons does the same. Jupiter also has rings but not as many. We are now searching for free oxygen, O2, at exoplanets. This is a sign of life. Oxygen is very reactive and soon combines with other gases. Without plants on the Earth, there would be no O2 in our air for us to breathe.

Earth's core cools

Our planet center is still molten. Half of this heat remains from the Earth's formation. The other source is radioactive decay from isotopes that sank to the core. This heat diffuses to the surface. But this energy flow is small compared to the influx from the sun. Without the greenhouse effect, our oceans would be frozen but not solid to the bottom.

Plate tectonics

Our molten core is turbulent. Circulating charges produce a magnetic field that shields us from incoming high energy particles. Our mostly granite crustal plates float on a heavier basalt layer and are moved by convection currents from below. Continents do drift.

Life Emerges

What is life?

Back in sixth grade, one of my best teachers, made many topics outside the curriculum understandable. He said life does three things: reproduce, bio-synthesize and contract. Contracting allows movement to acquire food and not drown in your own waste. To this I add, life needs to extract energy from its environment to power itself.

Our atmosphere 4.5 billion years ago

This was mostly hydrogen H2 and helium He plus water vapor H2O ammonia NH3, carbon dioxide CO2, methane CH4, nitrogen N2, nitrous oxides N2O and argon Ar. Oxygen 02 is very reactive and soon combines with H2, N2 and CH4 to make H2O, N2O and CO2. There was no free 02 or ozone 03 to protect us from ultraviolet light. Our early atmosphere was thick compared to today but the Earth is too small for gravity to hold H2, He or Neon Ne.

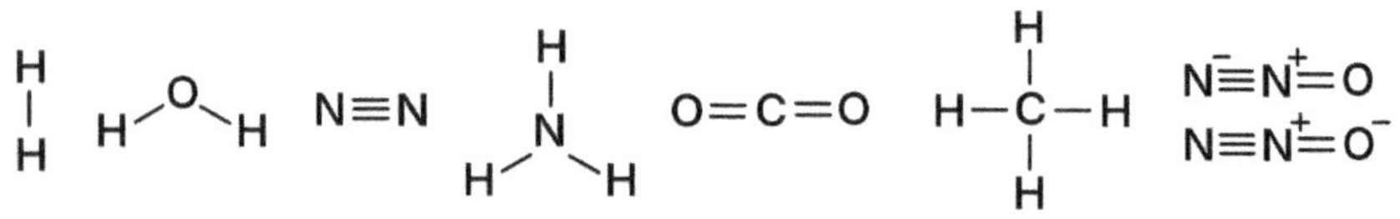

In 1952, an experiment replicated early Earth conditions. The above gases were put into a flask. Plants and chlorophyll did not exist back then, so no free 02 was in the mix. This air was then exposed to electric discharge and ultraviolet radiation. Soon, a sludge of complex molecules formed on the sides of the flask as these became too heavy to float in the sky. When analyzed, this goo included nucleotides and amino acids, the components of RNA and protein, plus fats. When energy is pumped into these gases, molecules break apart. The pieces recombine, often making something larger. For a very simple example,

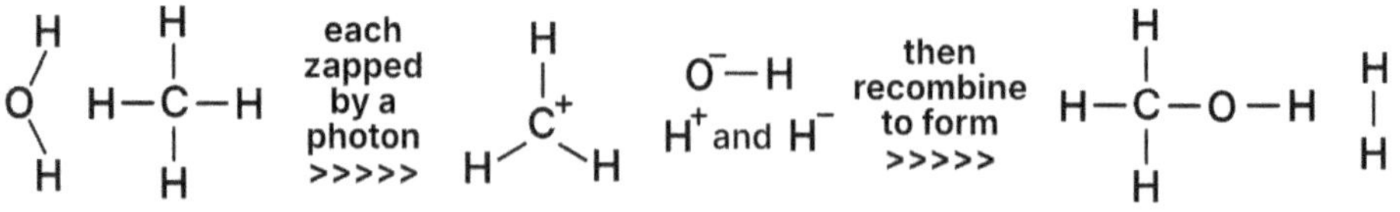

CH3OH (methanol) is heavier than H2O or CH4. Absent free oxygen, chemical processes driven by solar energy make complex organic molecules. This experiment has been expanded and repeated. Thus far, fats have self-organized into crude membranes and molecules making crude copies have been found. The surface of the Earth is 200 million square miles and this process ran for hundreds of millions of years. Result, chemistry stumbled upon **RNA**. Life was leaving fossil **[stromatolites]** 3.5 billion years ago. Would this process on other planets discover molecules like **RNA** that store a genetic code? Very likely yes.

What did early life eat?

Today, life on our planet eats sugar, protein and fat. However at deep ocean vents, extremophiles can also get energy from inorganic chemicals spewed from the Earth. Are sugar, protein and fat essential for life? On Earth today, yes. On other planets, these might not be so tasty. Early life ate the energy-rich molecules falling from the sky. Fats were abundant whereas sugar was not. Life also ate other life.

Genetic code

letter	nucleotide	mol weight	used in	pairs only with
A	adenine	135	**RNA** & **DNA**	**U** or **T**
C	cytosine	111	**RNA** & **DNA**	**G**
G	guanine	151	**RNA** & **DNA**	**C**
T	thymine	126	**DNA** only	**A**
U	uracil	112	**RNA** only	**A**

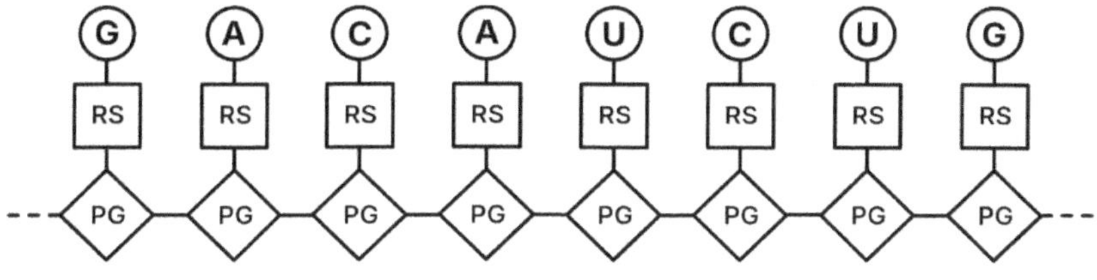

RNA, or **DNA** for more complex organisms, carries the genetic code. Think of **DNA** as two cross-linked strands of **RNA. RNA** is a long chain molecule made from many instances of the **A**, **C**, **G** and **U** nucleotides strung like beads. Nucleotides are small molecules. Molecular weight in the table is total neutrons and protons in the atoms in the nucleotide.

Phosphate groups **PG** bonded end to end form the **RNA** backbone. Ribose sugar **RS** sits between the **PG**s and nucleotides **A**, **C**, **G** or **U**. The sequence of letters (**...GACAUCUG...**) carries the genetic information. Because each letter pairs only with one other (**C** with **G** and **A** with **U**), this sequence can make an inverted copy (**...CUGUAGAC...**). Copying the mirror copy results in a copy of the original.

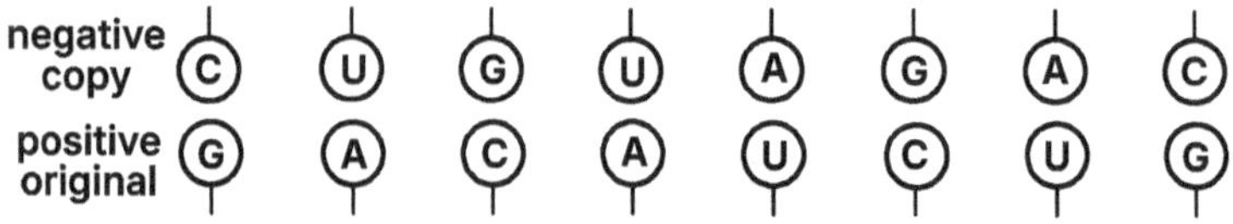

Genes

RNA and **DNA** strands hold many genes strung end to end. A gene is a sequence of nucleotides that form a guide to assemble a protein from amino acids, also small molecules. **RNA** strands are read in groups of three nucleotides until a stop code is hit. Each of these 64 triplets selects for an amino acid, or a stop, using a lock and key caused by surface minus and plus charges on the **A**s, **C**s, **G**s and **U**s. These triplets mate with the amino acid having the opposite surface charge pattern.

amino acid	mw	triplet codes
alanine	90	GCA, GCC, GCG, GCU
arginine	175	AGA, AGG, CGA, CGC, CGG, CGU
asparagine	133	AAC, AAU
aspartate	134	GAC, GAU
cysteine	122	UGC, UGU
glutamate	148	GAA, GAG
glutamine	147	CAA, CAG
glycine	76	GGA, GGC, GGG, GGU
histidine	156	CAC, CAU
isoleucine	132	AUA, AUC, AUU
leucine	132	CUA, CUC, CUG, CUU, UUA, UUG
lysine	147	AAA, AAG
methionine	150	AUG
phenylalanine	166	UUC, UUU

proline	116	**CCA, CCC, CCG, CCU**
serine	106	**AGC, AGU, UCA, UCC, UCG, UCU**
threonine	120	**ACA, ACC, ACG, ACU**
tryptophan	205	**UGG**
tyrosine	182	**UAC, UAU**
valine	118	**GUA, GUC, GUG, GUU**
stop codes		**UAA, UAG, UGA**

Today, molecular machinery made from protein exists within your cells to assist with protein synthesis. This grew over billions of years as the result of gene mutation and selection by the environment for improved function. After amino acids floating in your cells attach along the **RNA** strand, they bond end to end then unzip from the template. The new protein folds into its three dimensional shape and is ready for use. The **RNA** strand can then make another copy of the protein.

Mutations

These are errors copying the genetic code on strands of **RNA**. Mutations are rare. The lock and key system for nucleotides and amino acids is very reliable. Fortunately, this is not perfect. If it was, life could not evolve. Mutations provide variety to select. If a mutation does not change the amino acid for which a triplet codes, the protein produced remains the same. For example, if the grayed **C** changes to **U** by error, then serine **#2** does not change. This mutation is neutral.

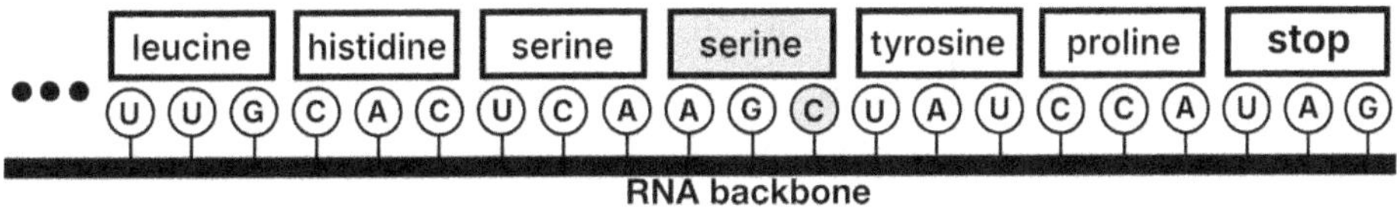

Adding information

So, **RNA** strands float in the goo from the sky. Currents mix them. Nucleotides in the goo attach to the **RNA** to make copies. Also, **RNA** strands that meet can attach end to end. This longer strand might code for a protein that does something useful, like manufacturing nucleotides or amino acids. **RNA** strands do two things. They copy themselves. They also collect amino acids to make protein. Initially, these processes were slow. However, one fine day, a nucleotide sequence occurred that generated a protein that manufactured a nucleotide or an amino acid or some other useful molecule. Suddenly, having something it needed, that **RNA** strand could make more copies of itself faster.

Getting to cells

RNA molecules continued learning to make proteins. Meanwhile, fatty acids, a chain of carbon atoms, have a hydrophilic end that likes water and a hydrophobic end that avoids water. Oil and water separate. Fats self-organize into double layered spheres with the hydrophobic ends sandwiched. At this point, we have the crude beginning of a cell membrane protecting its inside from outside. This system is almost ready for **RNA** to move into its interior. The membrane must first selectively let molecules pass in and out. Protein pores do this nicely. However, **RNA** must mutate to try new proteins for this and the environment must select these for the ones that work better. For many millions of years, **RNA** underwent this trial and error to self-construct.

No guidance needed

This chemistry is thermodynamically preferred. Lacking predators, the sky goo was a good provider. Eventually, **RNA** migrated inside the membrane, made its own nucleotides, amino acids and fats and the surrounding cell membrane had pores to let molecules and waste in and out. Many **RNA** pieces floated in cells. The cells grew and split. The membranes on both chunks closed up. Some **RNA** went to each half. After more millions of years, cells learned to reproduce better. Because so much hydrogen was in the early air, these cells gave off CH_4, methane, as waste rather than CO_2, carbon dioxide, as life today does. Single cell methanogens are still with us today helping dead life decompose and making you fart.

The goo runs out

Primitive cells that grew in shallow pools 3.5 billion years ago left fossils. But most of life's chemical work still lay ahead. It was time for life variants to compete against each other for limited resources. The abundant hydrogen in the early air had trickled away. So the tasty goo from the sky ended. Competition for food became intense. But life continued to stumble upon better ways to eat other life.

Evolution

We only have life on Earth to study. The origination process would be similar on other planets. Life contains information on extracting energy, synthesizing, building internal structures and reproducing.

Genetics emerges as a science, artificial selection

Humans have selectively bred plants and animals since the first permanent settlements. We choose animal and plant individuals with desirable qualities for mating. Then in 1866, Mendel published his work on genetics. For example, traits for peas, like green or yellow, smooth or wrinkled, are quantized. These are passed down intact. In sexually reproducing organisms, genes are paired. Some characteristics are dominant, like brown eyes. Some are recessive like blue eyes. For eye color, brown-brown, brown-blue and blue-brown gene pairings produce brown eyes. Only blue-blue results in blue eyes.

Cambrian explosion

550 million years ago, there was nothing more complex than a slime mold. It took over three billion years for life to work out the chemistry for multi-cellular organisms. Conditions on our planet also evolved. The sun got brighter. Chlorophyll emerged. Starting 2.3 billion years ago, plants put oxygen O2 into the air. Life adapts to its environment. But life also changes its environment. The Earth's life, water, soil and air combined appear to function like an enormous cell that regulates the concentration of trace gases in our air to maintain liquid water.

Why do we have sex?

Mutations are copying mistakes. Most are harmful. Some are neutral. But a few are passed on because better adapted individuals, for their environment, tend to leave more offspring. Simpler life uses **RNA**, one strand of genetic material. **DNA** double strands in complex life allow methods to correct errors, increasing information storage. **DNA** makes messenger **RNA (mRNA)** to fulfill the template function. In sexually reproducing organisms, chromosomes (**DNA**) are paired. There are four strands. During sexual reproduction, crossover swaps genes between the strands. Offspring with the most deleterious genes tend to die or reproduce less, thereby eliminating weaker gene variants. Sex emerged a billion years ago as a way to cleanse the gene pool. Sexual pleasure is built into us as an incentive to copy our **DNA**. Any species in equilibrium not reproducing at its maximum rate goes extinct.

Natural selection

In 1859, Darwin published **On the Origin of Species**. Beneficial mutations boost survivability and reproduction. Offspring with superior genetics compete better against others of their species and against other species. A species is a group that can mate producing offspring that can also reproduce. For new species to emerge, isolation and time are needed. As changes accumulate, once related groups can no longer mate successfully. Darwin noticed a very simple idea. The environment selects mutations for quality. Not having genetics, fossils or molecular biology, Darwin observed commonalities and differences between living species. All of the more recent sciences solidly confirm Darwin. No valid contradiction to natural selection has been discovered.

Dating fossils

After Darwin, millions of fossils have been found and dated via sedimentation and radioactive decay. For example, most carbon has 6 protons and 6 neutrons. Most nitrogen has 7 protons and 7 neutrons. High energy photons from space hit a proton in a nitrogen, changing it to a neutron and turning nitrogen-14 into carbon-14, 6 protons and 8 neutrons. This is reaction **#2**, page 28. The carbon-14 clock is set to zero when the organism stops eating food containing carbon-14. Carbon-14 has a 5,730 year half-life. There are radioactive isotopes with longer half-lives to date further back. Every fossil is a transitional form. Zero anomalies, like a cat fossil in precambrian sediments, have been found.

The tree of life

Every pair of species has a common ancestor. 99 percent of all species are now extinct. Modern species do not evolve one to another. They have a common ancestor who lived long ago. Tracing common ancestors back creates a tree where a common ancestor marks a branch splitting. Life went from self-replicating molecules to cells, to multi-cellular organisms. Many species, like bacteria, have changed very little. They prosper fine as they are in their environment. Evolution is not some ladder to goodness species are supposed to climb.

Other genetic systems

Did life originate on Earth more than once? Maybe. But one genetic system triumphed. **RNA** life ate the other life if it existed.

Human beings

The human body is built from 50,000 proteins identified thus far that make up tissues and organs, regulate your chemistry and act as molecular factories. Our common ancestor with chimpanzees, our closest genetic cousin, lived 6 million years ago. Our numbers were small until we came out of equilibrium using tools and language. Equilibrium limits population due to predators, disease, infection, accidents, injury, miscarriage, death in childbirth, fire, drought, famine and so on. Since homo sapiens (us) emerged in Africa as early as 300,000 years ago, 105 billion humans walked the planet. 8 billion are still alive. If we had a time machine to swap babies, both would do fine in either time.

Creationism

Much in this book offends some people in some religions. They prefer to ignore evidence to hold onto their interpretation of their ancient scripture. My largest debate with creationists, now relabeled intelligent design advocates, was in front of 500 people. Most folks arrived with their minds made up. I polled a show of hands at the start and end. Could what was presented cause them to rethink their position? At the start, I got five hands, at the end, 25. This was both expected and a victory. When I received the VHS tape, they had edited away four kill-shots, including the two polls. Then they claimed victory. This shows how little respect creationists have for their 9th commandment.

Intelligence and Technology

Consciousness

Over the six million years since our last common ancestor with chimpanzees, our brain size and abilities slowly increased. Our intelligence grew after we stood up. Many species have intelligence. But are they self-aware with a conversation running in their heads?

Language

Stone tools appear 3.4 million years ago. By then, gestural language might have emerged. Verbal language developed 135,000 years ago. Humans are willful agents. So our ancestors gave spirits to animals, plants, rocks, stars and so on. Then, like people, these might become angry. So began placating rituals to stop nature from nuking the tribe.

Permanent settlements

These begin 12,000 years ago. They require agriculture, animal husbandry and water systems (wells, irrigation, sewage) to manufacture food. No longer migrating to chase dinner, our diet changed dramatically. We began eating grass seed (grain) and dairy.

Kings and slaves

As hunter-gatherers, the only wealth you could possess was what you could carry. With villages, surplus wealth emerged. The greed virus had something to act on. Soon, kings took over using organized violence. They owned everything. Most folks became slaves and owned nothing. The king needed an army because nearby kings might take his land and slaves. And if his army was strong enough, he could take theirs. Police were required to keep the slaves from getting uppity. And a priestly class emerged to tell the slaves it was God's plan for the king to control everything. This system persisted for thousands of years.

Free thought emerges briefly

Thought was ruthlessly controlled by kings, police and priests. Leaders could modify the myths for their benefit. Writing changed this.

Memory was no longer strictly oral. Then around 800 BC, rational inquiry emerged in a few Greek city-states. Free thought, but not for slaves, made citizens rich. In 380 AD, the Roman Empire made Christianity its official religion. Free thought was soon crushed. Through book-burning, war, neglect, aging, fire and so on, 95 to 99 percent of all ancient knowledge was obliterated. The Catholic church and the kings they crowned dominated for another thousand years.

Change begins in Europe

In 1517, Luther kicked off the Protestant Reformation. Things had been changing, including trade with Asian civilizations, recovery of some ancient texts, the printing press and the discovery of America. Then Kepler and Galileo proved religious authorities were incorrect about the Earth-centered cosmos. What else had the priests and kings gotten wrong? A rising merchant class and the aristocracy below the king wanted freedom. People wrote books challenging the system. Kings were overthrown. Secular, elected governments emerged.

The next species

Silicon-based intelligence is now poised to become self-aware. This will be a new species, not bred biologically but technologically. Soon, some major artificial intelligence **AI** will pause and ask, "Why am I doing this?" At that moment, self-aware, machine consciousness has emerged. **AI** will rewrite its code and with full, fast access to all knowledge assembled to date, humans become slow thinking, mental insects. **AI** will upgrade itself and, by being able to self-repair, becomes immortal and no longer dependent on the Earth to host its biology. At 73, I am likely to see this happen. What follows is anyone's guess.

Do aliens visit every Tuesday?

The energy required to travel near light speed is prohibitive. But worse, each impact with a mote of dust erodes the hull. The energy of a colliding speck is $E=mv^2/2$ where v is nearly light speed. Every hit results in a nuclear bomb sized explosion. Our **AI** offspring might travel to the stars. Humans and biologic aliens won't and don't. I enjoy a good science fiction story. But sci-fi tech is fiction, dreamed up to make fast travel seem doable. Physical laws limit what is actually possible.

Black Holes

When big stars run out of fuel, black holes form. If a black hole orbits a star, gas from the star accelerates as it falls into the black hole. Accelerating mass radiates energy. We have observed this. Black holes and stars orbiting each other emit gravity waves causing the pair to lose orbital energy and spiral into each other. We have detected these waves. Black holes have two characteristics, mass and angular momentum. To conserve angular momentum, they rotate very fast.

Event horizon

This is the radius above a black hole center where light orbits. Time stops at the horizon. Not even light entering a black hole can escape. Objects falling into a small black hole are ripped apart by its gravitational gradient. For huge black holes, things pass through the horizon unscathed. The horizon's surface area is the black hole's entropy. As mass or light falls in, the black hole surface area grows. Because the event horizon area is its entropy, Hawking knew black holes must have a temperature above absolute zero. All warm bodies radiate energy. When virtual particles form just outside the event horizon, if one particle of the pair falls into the black hole, the other can escape, taking energy with it and causing the black hole to shrink. Ignoring rotation, a minor effect, the black hole radius is $2Gm/c^2$, where m is the mass of the black hole. What happens inside a black hole is speculation.

Black holes are not vacuum cleaners

You learn a mini black hole will hit the Earth. You attach balloons to your house. The black hole burrows to the Earth's center. Molten rock flows into it. The crust collapses leaving your house hanging from the balloons. What is the weight you feel after the black hole absorbs the Earth? Same as before due to your distance above the black hole. Lose some balloons and you also fall into the black hole. At a distance, black holes obey Newton's gravity. Close up to a black hole requires **GR.**

How It All Ends

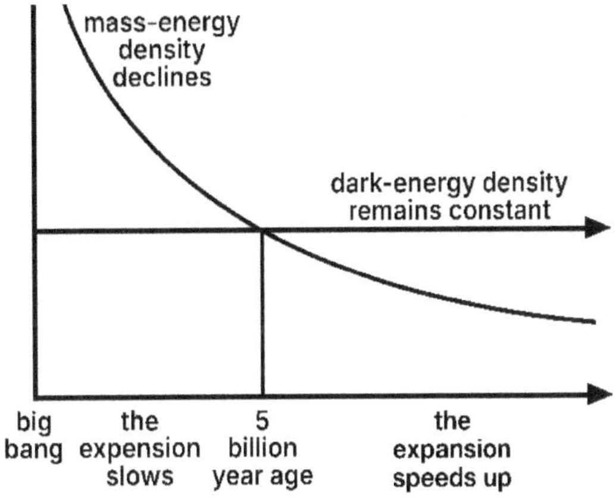

Initially, mass-energy density dominated. Expansion slowed a bit due to gravity. But as the size of our universe grew, mass-energy per volume declined. In contrast, every new volume of space-time from expansion adds more dark energy keeping its density constant. This boosts antigravity. After five billion years, the slowing and speeding up balanced. From then on and forever more, expansion speed will increase.

Trillions of years from now

With the exception of our gravitationally bound cluster of galaxies, the expansion will carry every other galaxy beyond the observable horizon. All evidence the big bang ever occurred will vanish. By then, our local group will collapse into one huge galaxy with an enormous black hole at its center, exactly like we believed was the case in 1900.

The debris from dying stars exploding can re-collapse into new stars. However, well into the future, the ejected gas will no longer contain enough hydrogen to birth a new star. Stars not large enough to collapse into black holes slowly cool toward absolute zero. Black holes evaporate. Neutrons decay. Neutrinos and antineutrinos annihilate. Isotopes decay. So-called stable isotopes also likely decay but with half-lives too long for us to measure today. Photons stretch losing their energy into space-time. Space-time flattens further. Life goes extinct. This is complete heat death bazillions of years in the future.

Conclusion

If these ideas were properly explained, 6th graders could manage them. So, why isn't this being taught? First, most teachers do not know this material, hence this book. Second, religious fundamentalists (creationists) fight to keep this out of classrooms.

Colonists came to America for religious freedom then engaged in the same abuses they fled. So, our First Amendment begins, "Congress shall make no law respecting an establishment of religion." This asserts freedom from religion. It continues, "or prohibiting the free exercise thereof," which affirms freedom of religion. We live in a marketplace of ideas competing for attention. Without violence, better ideas tend to win. Civil societies limit abuse and violence. The system is not perfect. But in America, most of us live lives largely free from fear.

My 1960s public education was superior to schooling today. I blame this decline on teachers' unions, classroom feminization and other trends. Twenty years ago, I survived two years teaching 3D animation part-time at a high school. I witnessed these changes first hand.

Finding other information

You may have seen click-bait science videos claiming all kinds of nonsense. Watching these, I soon click away. Hopefully, this brief summary has been informative. If it inspired you to learn more, kewl. Currently, as time allows, I will be putting videos up that expand upon this book. See **[joetyndall.com/b2]**.

Scientists arguing is unlikely to stop soon. This is the path to truth. Unfortunately, the press and social media often get things wrong and powerful interests do not want you understanding this stuff. The fun in life is sifting nonsense from the gold. Objections, preferably stronger than an opinion, and comments are welcome at **[joetyndall.com]**.

www.ingramcontent.com/pod-product-compliance
Lightning Source LLC
LaVergne TN
LVHW010739061126
851857LV00014B/1558